W9-BFA-330

WAR GAMES!

Dutch Leopard 1 MBTs in a field firing phase of 'Certain Strike'. Training with live ammunition is essential, although expensive, and requires strict safety controls.

WAR GAMES!

Rehearsal for Battle

ARNOLD MEISNER AND WILL FOWLER

ARMS AND ARMOUR

First published in Great Britain in 1988 by
Arms and Armour Press, Artillery House, Artillery Row,
London SW1P 1RT

Distributed in the USA by Sterling Publishing Co. Inc.,
2 Park Avenue, New York, NY 10016.

Distributed in Australia by Capricorn Link (Australia) Pty.
Ltd., P.O. Box 665, Lane Cove, New South Wales 2066,
Australia.

© Arnold Meisner and Will Fowler, 1988
All rights reserved. No part of this book may be reproduced or transmitted in any form or by any means electronic or mechanical including photocopying recording or any information storage and retrieval system without permission in writing from the Publisher.

British Library Cataloguing in Publication Data:
Meisner, Arnold
War games!: rehearsal for battle.
1. War games – Manuals
I. Title II. Fowler, William, *1947–*
793'.9
ISBN 0-85368-947-4

Designed and edited by DAG Publications Ltd.
Designed and edited by David Gibbons; typeset by Typesetters (Birmingham) Ltd.; camerawork by M&E Reproductions, North Fambridge, Essex; printed and bound in Portugal by Printer Portuguesa

Glossary

This is a list of the abbreviations and technical terms that may be encountered in British or US Army exercises. NATO standardization has produced a reasonably high level of commonality but there are still initials that may have dramatically different meanings according to the army in which they are used.

AAC	Army Air Corps
AAR	After Action Review
ACC	Area Co-ordination Centre
AD	Air Defence
ADC-T	Assistant Division Commander for Training
AFCENT	Allied Forces Central Europe
AFV	Armoured Fighting Vehicle
AMF	Allied Mobile Force
APC	Armoured Personnel Carrier
ARV	Armoured Recovery Vehicle
ATGM/W	Anti-Tank Guided Missile/ Weapon
AVKP	Armoured Vehicle Kill Potential
AVLB	Armoured Vehicle Launched Bridge
BAOR	British Army of the Rhine
Bde	Brigade
BMP	Boyevaya Mashina Pekhoty (Soviet APC)
Bn	Battalion
CALFEX	Combined Arms Live Fire Exercise
Cas	Casualty
CATRADA	Combined Arms Training and Doctrine Agency
CBW	Chemical Biological Warfare
CENTAG	Central Army Group
CPX	Command Post Exercise
Dam Con	Damage Control
DS/DISTAFF	Directing Staff
DD	Defence in Depth
Div	Division
DZ	Drop Zone
EC	Exercise Control
ECM	Electronic Counter Measures
EHIK	External Hit Internal Kill

▼ How it all began – a long time ago. British officers of a cavalry regiment in India at the turn of the century receive orders for an exercise. One group have white bands around their pith helmets to distinguish them as the 'enemy'. Exercises, as manoeuvres, have been part of military life for centuries.

ETR	Electronic Target Range
EW	Electronic Warfare
FAC	Forward Air Controller
FD	Forward Defence
FF	Field Firing (Live firing against targets in open country)
FFT	Force on Force Training
FGA	Fighter Ground Attack
FIBUA	Fighting In Built Up Areas
FOFA	Follow On Forces Attack
FROG	Free Rocket Over Ground
FTX	Field Training Exercises
GDP	General Defence Plan
GPMG	General Purpose Machine Gun
GSFG	Group of Soviet Forces in Germany
IGB	Inner German Border
KP	Key Point
LO	Liaison Officer
LFT	Live Fire Training
MBT	Main Battle Tank
MILES	Multiple Integrated Laser Engagement System
MOUT	Military Operations in Urban Terrain
MP	Military Police
MRD	Motorized Rifle Division
NATO	North Atlantic Treaty Organization
NBC	Nuclear, Biological and Chemical
NORTHAG	Northern Army Group
OC	Observer Controller or Officer Commanding
OP	Observation Post
OPFOR	Opposing Forces
OPSEC	Operational Security
POMCUS	Pre-Positional Material Configured to Unit Sets
PRA	Permanent Restricted Area
PUE	Pre-Positional Unit Equipment
Recce	Reconnaissance/ Reconnoitre
SAM	Surface-to-Air missile
SACEUR	Supreme Allied Commander Europe
SAW	Squad Automatic Weapon
SAWS	Small Arms Weapons Effects Simulator
SOXMIS	Soviet Military Liaison Mission
S-S	Surface-to-Surface
TA	Territorial Army
TAUT	Target Area Umpire Team
TEWT	Tactical Exercise Without Troops
TRA	Temporary Restricted Areas
UCC	Umpire Control Centre

CONTENTS

2A2-5ADA
B 23

THE SOLDIERS were ready for the helicopters. They knelt in line, rifles in their left hands, facing the LZ. The unmistakable beat of the rotor blades of the Hueys was audible long before the helicopters were in sight. First one appeared over the treeline, then another – and finally six flared into the open field, kicking up dust and stubble. There was a brief pause as officers and helicopter crews conferred and then the mad dash across the dusty soil to the bench seats on either side of the helicopter. The engines started up with a whine; the rotors changed pitch, and for a moment the Hueys hung about five feet off the ground. Then simultaneously they swung right and lifted off.

We travelled at treetop height – low enough for the down-wash of the rotors to send the leafy branches thrashing about. Cows careered about the fields, and a single rabbit made a mad rush to the cover of a hedge. The helicopters flew line-astern like a crazy roller coaster, hopping over the power lines and banking with a noisy 'whop, whop, whop' as the rotors cut the air. To our front and flanks gunships rode in armed escort.

I was sandwiched between a sergeant and a young soldier. The sergeant had seen it before: if he was excited it was concealed behind the quiet set of his face. But the young soldier hunched beside the door-mounted M60 machine-gun had caught the exhilaration of the ride. He yelled with spontaneous enthusiasm, a crazy 'war whoop' of delight.

For anyone who has flown in the noisy claustrophobia of a Wessex or a Puma, a Huey offers a unique ride – only the open doored little Scout helicopter approaches the experience. To imagine it, picture a narrow canvas bench; but where your feet would rest on the ground, they are braced instead on the alloy floor of the helicopter. Only inches away it seems is the rushing countryside 30 feet below.

We reached the LZ, and the Hueys came in line-astern. Above us the gunships flew in sweeping patterns covering the countryside. I suddenly wondered what the 1st Cavalry SOPs were for air assaults. Did they move away from the helicopter when it had landed and take up all-round defence, or did they group close-in around it? Would it keep 'burning and turning' or switch off on the LZ? We doubled away and took up all-round defence. The helicopters lifted off and roared over us at head height. Suddenly it was quiet. The company moved off the exposed LZ towards the treeline. We paused to orientate ourselves and then moved towards our objective, the bridge.

Within two or three hours we had secured the bridge and faced the prospect of either cold Meals Ready to Eat, or the local hospitality of a German gasthaus. *We were not at war. The air assault was part of the annual war games held by NATO, in this case 1/5 Cavalry recapturing the* élan *of the Air Cavalry of twenty years ago – but in 1987 in north Germany, not the highlands of Vietnam.*

◀ Air defence with a Vulcan gun system. This can be given a laser designator, but it is harder to operate. However, the role of air defence is critical in war and must be accurately simulated.

A day in the life of a British air-portable infantry company on Exercis

'Stand to' wakes soldiers just before dawn. They crawl out of their sleeping-bags, pull on combat jackets and tighten the laces on the boots they have been sleeping in. If the exercise has moved to 'NBC medium' they will have slept in their NBC suits; if to 'NBC high' they will have tried it in their respirators. For the minutes between darkness and the half-light of dawn they occupy their 'stand to' position. This is part of the command or defence layout of the position their unit holds. In the defensive it will be properly dug trenches; on the offensive it will be shell scrapes.

After they receive the order 'stand down', there may be time to wash, shave and cook breakfast. This is done in pairs so that one man is dressed and armed while the other is cleaning or cooking. The parapet of a trench makes an excellent breakfast table if the weather is good. On the offensive, however, there will be little time for domestic activity if a dawn attack or move before first light is required. Under these circumstances there is time to roll up the sleeping-bag, perhaps grab a cup of coffee or tea and then after Orders the vehicles start up, the infantry mount up and they are on the move.

Orders in most armies follow a strict formula. This has the advantage that tired brains can grasp them while ignoring those sections that are not relevant. As they move down the line from the general to the section corporal, they become more detailed and localized. The British divide theirs into: Ground, Situation, Mission, Execution, Administration, Command and Signals. Within this there are subdivisions as each unit receives its own detailed task. The Mission is repeated twice so that it is clear to everyone. A good officer or NCO will pause after giving orders to allow his men to absorb them, and then in turn will ask each man if he has any questions.

Orders may be preceded by a 'Warning Order'. This normally contains the phrase 'no move before' and is intended to alert soldiers that they will have work to do and that orders will be following soon.

On exercise troops may move out and be on the road just before the German commuters, but they will meet them soon afterwards early in the morning. Tanks, APCs and trucks will become mixed in with Mercedes, Audis and Volkswagens until the military vehicles have found some open fields or the civilian cars have reached their offices. The road-march may take up most of the morning unless there is an attack at first light. Tracked vehicles in the dark can be a hazard to men sleeping on the ground, so control has to be quite tight. The dawn attack, which is the outcome of the Orders received by the unit, will probably attract an audience. The local villagers will turn out to watch. The defenders will have stood to, and so unless they are very lazy or inefficient will not be caught by surprise. A rattle of blanks, bursts of GPMG fire, coloured smoke drifting in the trees, the thump of thunderflashes and shouting will indicate that the attack has 'gone in'. In war there would have been supporting mortar or artillery fire which would have 'neutralized' the enemy position prior to the assault.

After the attack the umpires will allocate losses. A supporting tank may find itself sitting off the road 'dead' for thirty minutes. For armour and infantry this is a chance to 'get a brew going' making hot tea or coffee. Men may check their kit, read a book or letter, or more likely just settle into the grass, half alert, waiting for the shout, 'OK, on your feet!'

Lunch time arrives. In a dug-in position this will be the snack section of a Compo ration, or an MRE pack. Like children, soldiers are inclined to pick at their food, eating the snacks they enjoy. In the afternoon orders may have come through for patrols in the coming night. The platoon commander will choose his men and they will study the map, rehearse the 'Immediate Action' or IAs on lights, obstacles or an ambush. The patrol commander will prepare his orders, and men will try to get some sleep.

During the day messages come over the radio and 'sitreps' and 'persreps' are sent by the company second-in-command. The sitrep is a 'situation report', in essence: 'This is where I am, and this is what I am doing.' The 'personnel report' gives the number of men in the unit; a 'logrep' is a logistics report. Persreps and logreps inform the B echelon about the numbers and needs of the unit. From echelon comes 'replen' or replenishment food, water and ammunition. On the move replens can be very difficult as vehicles become separated, map references are misread, or simply good camouflage makes the unit hard to find. Men are always hungry, and corner stores in German towns are quite used to men in NBC clothing, helmets, rifles and webbing joining in the queue at the checkout with a basket of shopping. The temptation of a beer in a *Gasthaus* can be very difficult to resist; only the US Army is 'dry' on exercise.

Air attack may neutralize the vehicles on the move, so once again it is time to stop at a roadside and a wait while the umpires decide the outcome of the attack. The attack will have been dramatic and noisy, as fighter ground-attack aircraft roar low over the vehicles.

The company commander will have left the company under the command of the 2IC (second in command) as he goes to receive orders from the colonel. He will return in the late afternoon to pass them on to his platoon commanders. The colonel may have had his orders from the brigadier or may have received them from the battalion liaison officer who has shuttled between the brigade and battalion HQs.

Evening draws in, the night patrol has checked its kit at the defensive position and is ready to move out to its objective. Sometimes there really is something there – vehicles and men;

sometimes the patrol may have been set up in order to keep men busy. They may guess this but go through the motions correctly. Night patrols in a peacetime Germany give curious insights into German domestic life as uncurtained kitchen windows show a glimpse of a tidy, well ordered world. In the dark, men with faces smeared with camouflage cream and rifles and equipment taped so that they do not rattle move silently through the fields at the bottom of the garden. The leader has his map and compass, but the road pattern and street lights give a pretty reliable guide. Crossing roads at night takes on the stealth and desperate energy of a nocturnal rodent – a sprint from one drainage ditch or hedge to another.

If men are living off 24-hour one-man packs they will have cooked their food before it has become dark. If there is central feeding the food will have come to the position in insulated food containers, hay boxes or Norwegian containers. This type of feeding is done after dark and men move by half-sections down to the containers to fill their mess tins with stew and hot pudding. The food is never *cordon bleu*, but it is hot and it breaks up the day. Ammunition and water will come with the meal as will letters and newspapers. There will be a chance to buy cigarettes and sweets from the colour sergeant who has brought them with the rations. Refuse, in black plastic bags, will be removed.

On the move, vehicles will not stop for the night until last light. Then they will pull into a field that has already been harvested and so only stubble and earth will be rutted by wheels and tracks.

Night routine will include a guard duty and the allocation of arcs of responsibility to men and weapons. The unit will be in 'all-round defence'. Sleeping-bags will be unrolled and 'bashas' (shelters made from waterproof ponchos) will be erected. A latrine area will have been dug and, before stand to, men will have used it or simply the hedgerow, depending on their needs.

Silence will follow 'stand to' as day slides into night. After that, apart from whispers between guards or the brief violence of a night attack, it will remain silent until the dawn.

'Certain Strike'

Although this book is not intended to be 'the book of the exercise', Exercise 'Certain Strike' does provide the basis for many of the illustrations and explanations. The 'Orbat' (Order of Battle) for 'Certain Strike' gives an idea of how this Reforger FTX was structured and how the different national contingents were joined by men from the US III Corps. The Staging Area, through the deployment Area to the FTX Area is shown on the map. It spread almost completely across West Germany from the garrison towns of Münster and Osnabruck to the edges of the Inner German Border. The Deployment Area was comparatively narrow, with limited roads, so the movement of men, vehicles and helicopters had to be controlled carefully. In the FTX Area is the large tract of land in the Soltau Hohne training area, which was used for the 'tank battles'. In addition to men on the ground, a Corps HQ CPX was grafted on to the exercise, in which German, British, Dutch and Belgian headquarters were able to fight a 'map war', albeit on the back of an FTX.

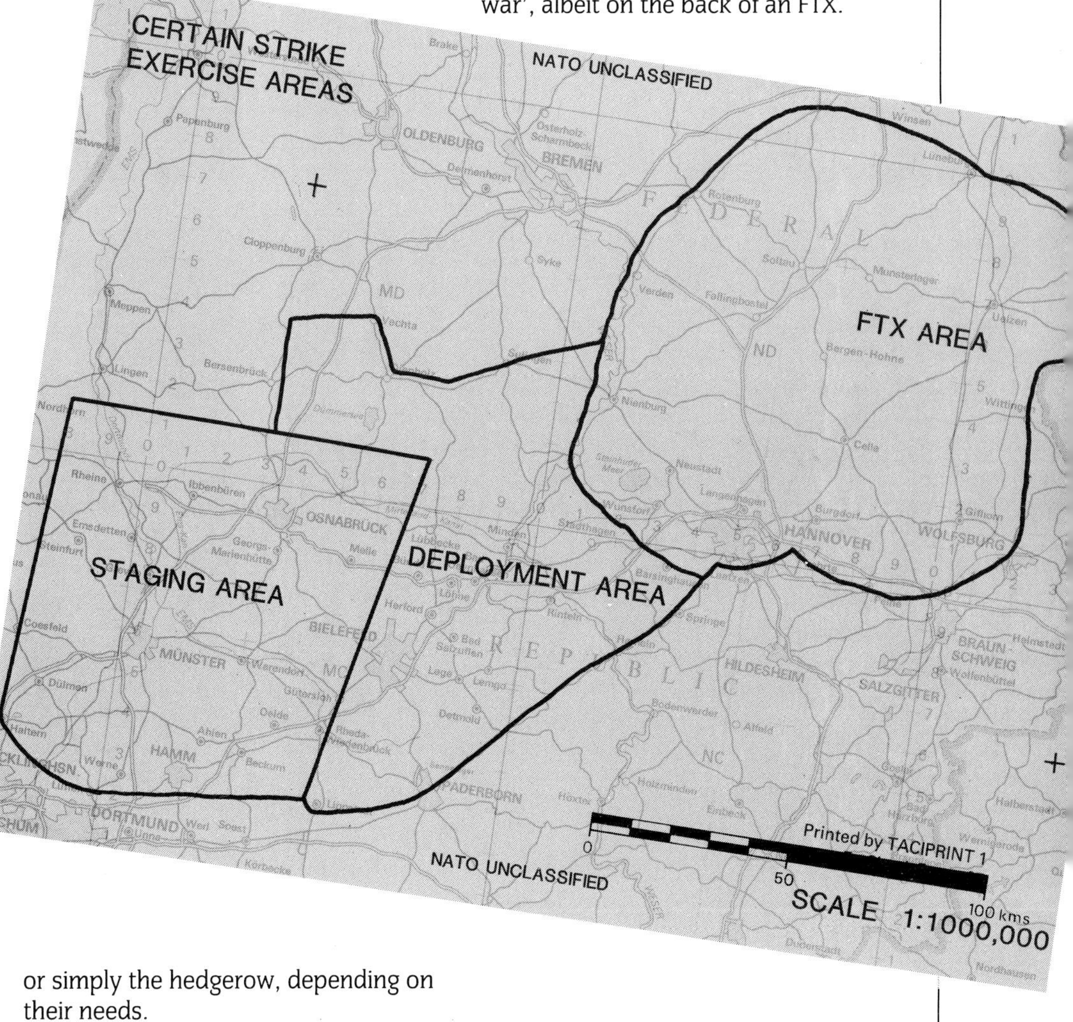

IT IS HOPED that this book will give the reader an informed insight into the workings of a major field training exercise. Exercise 'Certain Strike' was held in 1987 in the Northern Army Group area of West Germany. This was the biggest exercise of its kind since the Second World War and involved some 78,300 soldiers, among them the US III Corps, consisting of the 1st Cavalry Division, 2nd Armored Division, 6th Cavalry Brigade (Air Combat), 45th Infantry Brigade (separate) as well as corps troops and HQs. The US reinforcement of troops to Europe was an incredible logistical undertaking. The American photographer Arnold Meisner followed the US soldiers from Fort Hood, Texas, to the Netherlands, where they joined their vehicles and equipment following the phase designated 'Reforger'. Much of this hardware had been shipped over especially and held in stores in Germany, Holland and Belgium. I joined the exercise later in Germany during the 'Certain Strike' element, and there I spent time with US units in the field. As both a reserve soldier and a defence journalist, I had previously exercised with the British Army, so I was interested to see how the US Army operated. My observations of 'Certain Strike' and recollections of exercises stretching back into the 1970s provide the basis for the book, while Arnold Meisner's photographs follow the action.

The chapters are 'free-standing' but allow the reader to build up a picture of why FTXs are held, how they are made to work and where the important front-line areas are. The future of such training is also examined as sophisticated laser systems begin to operate – these allow 'kills' to be made accurately for the first time, thereby removing the element of uncertainty that has plagued exercises for decades.

War Games is not a history of training exercises nor a guide to paint-pellet airgun pastimes. It is an attempt to convey the reality and *un*reality of field training exercises, principally in West Germany but also in back street mock-ups, jungle trails, bitter snow and even the Mediterranean sun. Sometimes these 'games' can be very convincing; sometimes their fake nature imposes itself in a crazy and irritating way. Try lying camouflaged in a wet, muddy ditch on a Sunday morning as the elderly inhabitants of a small German village pass you by on their way to church. They smile at you because they have seen it all before. And when they were your age, they were caught up in the grim *reality* of war – not games.

Introduction

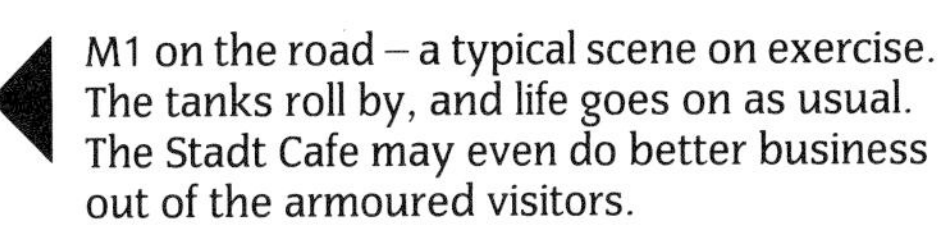
M1 on the road – a typical scene on exercise. The tanks roll by, and life goes on as usual. The Stadt Cafe may even do better business out of the armoured visitors.

There are a myriad of names today to describe the deployment of men and equipment from their depots and barracks into the countryside. Modern training techniques have always managed realistically to simulate most of what the combatants would experience in war: the briefings, identical movement logistics, camouflage and electronic warfare. Reality always imposed itself, however, when 'contact' was made and blank fire was exchanged between 'hostile' groups. Today even that has been effected by new technology. Games are fast becoming very similar to war itself.

'Exercise' is an imprecise term and can encompass the deployment of as few as 30 men with small-arms to many thousands equipped with aircraft, vehicles and helicopters. Their common theme, of course, is the acting-out of war – minus its death and destruction. Today's exercises are umpired by directing staff (DS) who 'score' the game and attempt to teach the participating soldiers tactical lessons. An exercise has to be distinguished from an *operation*, which is the actual employment of troops in war, and there have evolved a number of distinctive types.

Field Training Exercise (FTX)
This is the large-scale event that puts thousands of men and vehicles into the field and gets them as close as possible to 'war'. Their stamina and battlefield skills are tested to the full; the reliability and effectiveness of their hardware is evaluated; and the logistic and command structure efficiency is evaluated.

Command Post Exercise (CPX)
Involving only the headquarters and signallers, with a few drivers and general-duty men, this is designed to test and train the officers of the command structure by presenting them with realistic problems of the type that they would encounter in a real war situation. Enacted either 'in the field' or in special training establishments, the 'war' is fought over the airwaves by radio and telephone and can be a very demanding process: information needs to be speedily collated and assessed, maps marked, orders drafted and sent, messages enclosed and decoded – and so on. The CPX

How it all began. The end of the Second World War divided both Europe and Germany into different zones of influence. Here French troops enter Kehl passing a shop dummy rigged out as a figure of Hitler. The French retain a presence in Berlin as part of the Four Power Agreement on the City. They are also taking an interest in the military side of NATO.

Which side feels more threatened? Here Soviet reconnaissance troops – who, in the event of an attack, would spearhead the advance – check the terrain during an exercise. Both NATO and the Warsaw Pact assert that their exercises are defensive in character.

truly weeds-out those who cannot remain calm and effective in a 'crisis'.

Live Firing Training (LFT)

With safety of paramount importance, it is best that not too many men take part at any one time, but battalion-size exercises are quite common. It is as close to the atmosphere of war as one can go without injury and is fairly flexible in that targets may be engaged with live ammunition from either a defensive position or during an advance to contact. In the latter situation the troops will move across in tactical formations with live small-arms and supporting mortar, artillery or tank fire. They will engage 'enemy' targets, which can accurately be reported as 'destroyed'. UK training areas are suitable for infantry weapons exercises, but for safe utilization of larger weapons the vast empty spaces offered by Canada are far safer. It is an invaluable training experience actually to smell the cordite in the field and to have live shells pass closely overhead. Men learn to trust their arms, and their dependence on their comrades is fostered. The distinctive 'crack and thump' of small-arms fire or outgoing/incoming artillery is recognized – and the experience may prove to be lifesaving in a 'real' situation.

Tactical Exercise Without Troops (TEWT)

Although very similar in many respects to the CPX, the TEWT obliges the participants to work 'on the ground'. It involves the command element in either an attack or defence scenario, the officers being briefed and then walking around the position they are required to defend (although in an attack they will not normally be able to see the enemy position at close quarters) in order to make an appreciation prior to planning and issuing orders. The officers would in this situation assess the likely weapon arcs and ranges, discuss where supporting fire from artillery, etc., could be directed. A task might range from defending a bridge to mounting an attack on a dug-in position. In the latter, an officer would need to assess the enemy without attracting fire, and from his no doubt restricted vantage point, he would decide on a left- or right-flanking attack or even a straight up the middle 'gamble'. In either scenario, defence or attack, covering the ground afterwards can be very informative: previously unseen topographical features may become evident, and when standing in the enemy position you may notice flaws in your attack that you did not think existed.

In Britain the exercise system depends on the directing staff for its effectiveness. These umpires control the game, having designed the battle being fought. Keeping it constantly monitored, they adjudicate and take the final decision on rulings for any command problems that arise. Their omnipotence has led to the phrase 'DS solution' entering Army venacular as a statement that needs no further explanation since it is unquestionably correct!

These various forms are generally excellent ways of teaching and testing, and with periodic exceptions they are not overly heavy on manpower. Preparation is generally the key to any successful enterprise, and warfare is no exception to this historical truth. As long ago as 500BC the Chinese military philosopher Sun Tzu wrote in *Art of War*:

'The general who wins a battle makes many calculations in his temple ere the battle is fought. The general who loses a battle makes but few calculations beforehand. Thus do many calculations lead to victory, and few calculations to defeat: How much more do no calculations at all prove the way to defeat!'

Trucks on board the USNS *Algol* (T-AKR-287) as she berthed in Rotterdam as part of 'Reforger 87'. Many of the trucks have container shelters secured to their decks.

A US Army movements controller checks a train-load of M1 Abrams before it pulls out from Rotterdam docks.

WHY?

JUST AS no play will succeed without rehearsals, nor an athlete triumph in competitive sport without training, so practice is essential if war is to be waged victoriously. Armies have always drilled themselves and honed their expertise. Their ultimate raison d'être is to fight battles, and this would be a suicidal undertaking if they were unprepared and ill-equipped. 'Readiness' is a military watchword as old as time itself. Naturally, training for modern warfare is a great deal more involved and complex than that required for ancient encounters. However, the basic truism remains: most military victories are due to good training, drills, leadership and sound tactics.

Modern NATO exercises have a demonstrative intent as well as a training aspect. They are to deter potential aggressors in the USSR, and they imply an ability to pose a threat too. Today the task of exercises are manifold: to test tactics, logistics, strategy and operational doctrines, to ascertain leadership qualities and fighting abilities; to allow men to familiarize themselves with terrain, climate and equipment; to instil the military ethos of group identity, mutual support and loyalty; to ground-test the hardware in realistic combat conditions; to foster and control aggression; and, so far as NATO is concerned, they are to demonstrate that it can function both in

▲ 'Reforger 83'. Spec 5 Sharon Sparks of the 28th Transport Detachment checks off the registration numbers of vehicles being offloaded from *Cycnus* at Rotterdam.

▲ 'Reforger 87'. Four years later, the light pen and the computer code have replaced the pen and clipboard. A US servicewoman with the full set of web gear and Kevlar helmet checks vehicles.

Sergeant M. Uvarov and Private E. Zhuravlev exit their BMP during a training exercise. They are both members of the Leningrad Komsomol Guards Training Regiment. The BMP is very widely used in the armies of the Warsaw Pact and other friendly or non-aligned nations.

terms of individual arms and as an integrated whole.

Exercise 'Certain Strike 87' aimed to practise plans and procedures for the employment of the US III Corps in the Northern Army Group (NORTHAG) area. The Corps is a reserve force for NATO's Central Region and the present concept of operations envisages the commitment of a strong, mobile reserve to counter any enemy aggression in the Army Group area. III Corps is the most modern-equipped armoured formation in the world, and its transatlantic airlifting to Europe (which it would need to do in wartime) during 'Reforger' was the greatest logistical exercise seen since the Second World War. These two exercises were a significant demonstration of US rapid deployment capability and NATO's readiness to realise its commitments.

The exercise allowed maximum free play to the Corps and limited free play to 'Orange forces' (the enemy). Among the operations the Corps had to conduct were passage of lines (moving through a friendly division to engage the enemy), launching counter-attacks, undertaking river crossings and assault river crossings, defensive operations and employing the Corps Aviation Brigade.

A constant theme that underlies exercises and often conflicts with official doctrine, is the aversion of the professional soldier to fighting a purely defensive battle, the outcome of which can be no more than defeat or stalemate. NATO has had numerous doctrines, all of which are dependent upon political considerations of their acceptability to public opinion. For instance, the topography of West Germany makes the most sensible military option a Defence in Depth (DD), which entails withdrawing in the face of an attack, sacrificing important areas of West Germany,

▲ BMPs on the move in a Russian winter. The rugged country in the USSR has forced industry to produce vehicles that have a reputation for their good cross-country performance. They can be used in both the dust of a Russian summer and the extreme cold of the winter.

then containing the enemy before amassing forces for a counter-attack. (Naturally, this is unpopular with many Germans.) To ensure this would be operable in reality, it has to be tested: hence exercises and, most notably, the arrival of the substantial US reinforcements into Europe without which any NATO offensive strategy could not be undertaken. 'Reforger' and 'Certain Strike' demonstrated forcefully to the Warsaw Pact that NATO was fully capable of implementing its doctrine. Due to the political fallout from advocating DD, the officially espoused NATO doctrine is Forward Defence (FD) which is more robust in character and neatly institutes DD – but as far forward as possible to the border with East Germany. These ideas are presently complemented by Follow on Forces Attack (FOFA), a doctrine that matches the aggressive needs of soldiers with political 'defence only' needs; it envisages deep strikes into Warsaw Pact territory upon the outbreak of war, therefore giving DD and FD – but behind enemy lines!

The Soviets view these developments with considerable alarm. Naturally, their ideological disposition is to regard the Warsaw Treaty as embodying a concept of universal security while the North Atlantic Alliance is seen as reckless militarism. A Soviet News Agency (TASS) military consultant, Vladimir Chernyshov, examined the number and variety of recent NATO exercises and wrote: 'Over the past few years NATO war games have acquired unprecedented scale and can hardly be differentiated from real war preparations. Military exercises more and more openly test scenarios for launching and waging different wars – short and protracted, local and global, nuclear and conventional . . . to meet the challenge, military exercises include planning and effecting large mobilization, troop deployments and strategic landings of US troops in the overseas theatres of operations. Another aim is to simulate joint realism with other NATO countries of operations involving conventional, nuclear and chemical weapons.' Elsewhere Chernyshov complains that NATO and the US armed forces 'hardly stop even for a day . . . suffice it to cite such regular war games as "Global Shield" (covering half the planet), "Autumn Forge" (Western Europe), "Team Spirit" (the Far East), "Bright Star" (Egypt and the Horn of Africa), and "Big Pine" (Central America). The task of each is to test and perfect plans for the preparation, launching and waging of aggressive operations.'

This assessment is sound on its facts, but flawed in its conclusions. Given the imbalance of conventional forces in Europe, General Galvin, Supreme Allied Commander Europe (SACEUR), would have some problems waging a conventional war if for some insane reason his political masters were to order it. The programme of exercises is intended to test deployment of troops to locations where they could stop a Soviet or Warsaw Pact attack. The Warsaw Pact conduct similar exercises on their side of the border, and their scenario of enemy attack and friendly counter-attack is the mirror image of the 'war games' that take place in Europe.

The Soviet Navy and Air Force deploy around the world, and amphibious exercises have been held off the northern coast of Vietnam as well as in the Middle East and East Africa. The modest Norwegian battalion-sized exercises that I observed close to the Soviet border in northern Norway was a discreet affair compared with the naval and air activity that could be seen across the border. The Norwegians are at pains not to offend their neighbours, and no joint Allied exercises can take place in Finmark. Allied aircraft are not permitted to fly over Norwegian territory on missions extending east of longitude 24°E, nor are Allied naval vessels permitted to call at Norwegian ports east of the same line.

The 'Brothers in Arms' exercises held by the Warsaw Pact include amphibious operations on the Baltic coast, which NATO experts see as models for landings in Denmark. The Soviet Army sees major exercises as political gestures, just as NATO sees its exercises as part of the deterrence process. Even the Norwegian garrison of South Varanga in Finmark makes a contribution to deterrence: it trains to fight both as conventional infantry and then as guerrillas, using caches of food and ammunition hidden in the countryside.

Both sides subscribe to the proverb attributed to General Suvorov: 'Train hard – fight easy.' Arguably, a

East and West. US A-10s overfly a T-62 in the US Army training area, West Germany.

East and West. A Russian colonel talks to a young soldier from the US 1st Cavalry.

war in Europe against the forces of the Warsaw Pact would be the hardest operation in the world. If the firepower and numbers of the Warsaw Pact are compared with any other potential enemy, they dwarf it completely.

It is a supreme irony that, while exercises serve as preparations for war that can alarm those whom they are intended merely to deter, they can now also help to provide a basis for common understanding. At the end of the Second World War the Allies (which then included the USSR) agreed to establish missions in their respective areas of occupied Germany. As time passed and the two new German states emerged, so these missions changed their character from liaison to intelligence-gathering centres. On the Western side, the French, British and US Missions would be tasked with observing the exercises of the Group of Soviet Forces in Germany (GSFG). With three teams in the country, they could divide their work so that one would follow the air activity, another look for new vehicles and AFVs, and the third follow tactics and operating procedures. The Soviet mission has come to be known by the abbreviated name of SOXMIS. Strictly the missions had a limited area in which they could travel in East or West

Soviet Naval Infantry (Marines) of the Order of the Red Banner Black Sea Fleet march away from their tented camp during a training deployment ashore. Marines would be used to seize the Danish islands and block the entrance to the Baltic.

Germany – from their base to the headquarters of the host army or through recognized crossing points into East or West Germany or Berlin.

In practice they attempted to move around within East or West and to observe, photograph and note troop activities – particularly during the major exercises in the autumn. The exercise areas away from recognized training grounds are declared temporarily 'off limits' to SOXMIS vehicles during this period. British soldiers receive a SOXMIS briefing when they arrive in West Germany for exercises and also a small card showing the SOXMIS vehicle registration plate and giving details of what they should do if they encounter a SOXMIS car operating in restricted areas. In simple terms they should box-in the vehicle using two vehicles like Land Rovers but not attempt to restrain or arrest the Soviet personnel, and then they should telephone for the military police at Herford. US servicemen also receive a briefing about the Soviet mission – the SMLM in their area is based on Frankfurt, while that in the former French area operates out of Baden Baden. Like the British, they are instructed that if a mission vehicle appears in a Temporary Restricted Area (TRA) or Permanent Restricted Area (PRA) they are to detain it, within certain strict rules, or note its registration. In addition, the colour and make of the vehicle, the time and date of sighting, location (town, grid reference, autobahn), the direction of travel (if it is parked or moving), occupants (age, sex, type of clothing) and special equipment such as cameras and binoculars that may be in the vehicle. When attempting to detain a vehicle in a PRA or TRA, no force should be used. When the vehicle has been boxed-in, soldiers are required to 'show military courtesy and ask the occupants for identification; make sure that the Soviet

A Romanian CDE Observer at the FTX phase of 'Reforger 87'. Each side was equally intrigued with the other.

A Russian CDE observer listens as a Dutch officer explains a phase of 'Certain Strike' to a Canadian observer.

East German, Russian and Hungarian CDE Observers at 'Certain Strike 87'.

Colonel Jiri Divis of the Czechoslovakian Army makes a point during 'Certain Strike 87'. Most of the Warsaw Pact observers had a good knowledge of English, even if they decided not to demonstrate it.

vehicle has no way to get away by suddenly driving through a ditch or similar action which must be expected; secure the scene and direct traffic on highway to keep obstruction to a minimum and ensure that the military police are coming to assume responsibility'. Once a detention has been effected, soldiers must not 'interrogate or question Soviet personnel; open doors, or search the Soviet vehicle; tell the Soviets why they are detained; enter into arguments or allow the Soviet personnel to intimidate with threats; release the Soviet vehicle unless instructed by competent authority'.

Each side plays its mission role for real and will report any 'illegal' treatment they may receive. On one occasion a SOXMIS car was stopped by a military helicopter, which landed on the road blocking its way; unfortunately, when the rules were drawn up in 1945 there were hardly any helicopters around, so there was no provision for them in the rules – the SOXMIS car took a photograph of the helicopter and sent in a complaint. The possibility of a contact with a SOXMIS car does add spice to exercises, and there are tales of chases across country and SOXMIS vehicles bogging-

West German Military Police discuss the merits of their hand-held radio with an East German CDE observer. The contact, however small, between East and West helps foster confidence and reduce tension.

down in soft ground. Certainly the Soviet Mission vehicles – which appear to be standard West German civilian saloon cars, have high performance engines and rugged suspension; they also carry radios and sophisticated camera kit. There were reports of British soldiers attempting to stop them by standing in the road who have been quite deliberately driven at and badly injured.

After units have moved out of a vehicle laager or bivouac site, SOXMIS officers have been known to check the refuse to see what they can find. Exercise notes, message forms, even letters can give a good indication of the unit, its level of training, discipline and morale. The Soviet intelligence-gathering agencies are great collectors and have enough personnel to sift through the jetsam of an exercise to learn low-level intelligence.

Such activities are tolerated to a certain level. Like diplomatic embassies, each side feels that it can benefit from it more than they lose by it. If the unwritten limits are exceeded, however, things can turn nasty, as when the Soviets shot and killed an American agent caught in a secure area. Rare incidents like this apart, the Missions do contribute to keeping the peace in Europe, for such access allows each side to judge fairly accurately when an exercise has become preparation for war. Two recent examples indicate that operations for war can be disguised as exercises: the Soviet invasion of Czechoslovakia in 1968 and the Egyptian crossing of the Suez Canal in 1973 were both launched after 'exercises', when troops were therefore in the field and kitted-up. Any activity in Europe is immediately noted, just as the 1968 invasion was predicted by NATO intelligence, since equipment and ammunition needs to be stockpiled.

In recent years this international bridge-building has become fairly formalized, and in 1986 some 35 nations signed the Stockholm Document at the Conference on Security and Co-operation in Europe. Among them were the member states of NATO and the Warsaw Pact. Within the 'confidence-building measures' of the document was the provision that each side could send observers to witness exercises involving more than 13,000 men. They were present at 'Certain Strike' in 1987 in Germany and later at 'Purple Warrior' in Scotland. Most of the observers were

More than fifteen years ago in 1973, two men of C Company, 16/1 Infantry Brigade of the 1st Infantry Division, were photographed on exercise in Germany. Hugo Keindengon, the local farmer, had seen it all before – he probably even recognized the mechanism of the M60, since it was derived from a German wartime design.

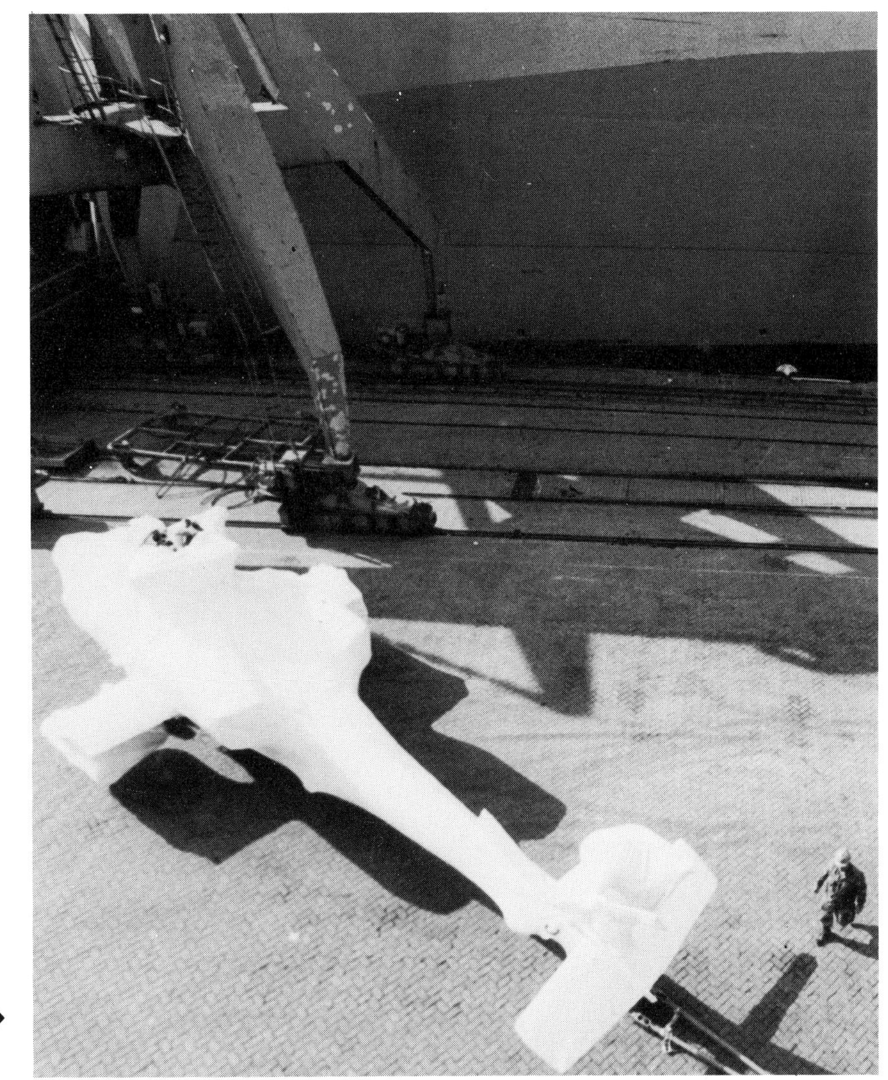

Still swathed in its waterproof protective covering, an AH-64 waits on the waterfront. Damage from salt corrosion on a long sea voyage is prevented by these covers.

An Apache AH-64 without its rotor blades after being offloaded from the USNS *Algol*. The arrival of these sophisticated attack helicopters caused considerable interest during 'Certain Strike'.

majors or lieutenant-colonels, and each country was restricted to two men. Accommodation was in a local hotel, and in Germany they were taken by coach to meet units and receive briefings. It was a unique opportunity to see the meeting of two groups of professionals, each slightly ill at ease in the political spotlight, but each intrigued to see how the other side solved problems. At a briefing by a West German brigadier, the East German colonel was both detailed and specific with his questions about tactics and planning. Although the officers spoke the same language and came from different parts of what had been the same country, they had different attitudes to training on exercise. The Western concept of 'free play' on an exercise was very hard to explain: clearly the East Germans planned their exercises to a low level and saw them as a way of testing operational drills, while the West Germans felt free to give their commanders some leeway on an exercise to test tactics and solve problems without the intervention of umpires or controllers. Free play had its political problems – anyone who has tried to track down a unit on the move in a major FTX will know how difficult this can be – especially if they are well camouflaged or sited. The Warsaw Pact observers clearly suspected that NATO was hiding men and equipment. In reality they had either moved from the grid reference at which they had previously been located, or were well camouflaged.

While governments and ideologies are often diametrically opposed, it is somewhat curious to observe the fact that armies everywhere are almost identical in structure and spirit. They may exercise differently and employ unique tactical methods, but soldiers and officers generally recognize their common bond – membership of a unique profession. Thus increased East-West military contact is in the interests of peace. After 'Purple Warrior', Colonel Georgi Gueorguiev of the Bulgarian Army said, 'How can we fight you now that we have met you?' His fellow observer, Colonel Mikhail Krutenko, a Soviet airborne officer, observed a drop by 500 men of 1 Para with great interest. Confidence-

▼ With weapons tagged, US soldiers wait in a sports hall at Fort Hood prior to boarding the flight for Amsterdam. Almost all service flights seem to involve waiting for orders after a very early rise.

Two soldiers check the information board as they secure their ALICE packs. Normally troops in transit carry basic essentials while their packs travel as cargo.

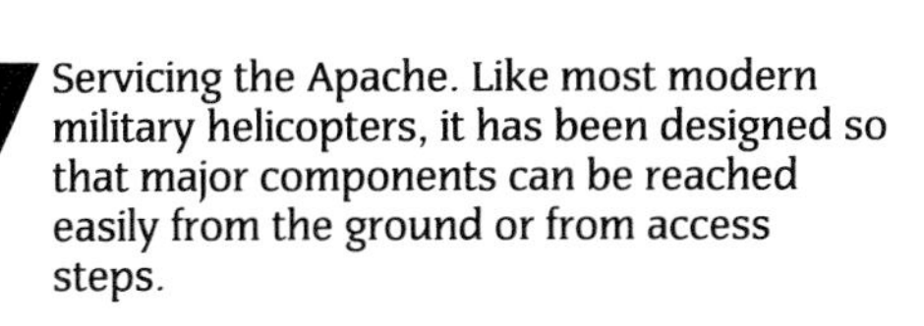

Servicing the Apache. Like most modern military helicopters, it has been designed so that major components can be reached easily from the ground or from access steps.

Men of III Corps board a DC-10 in the warmth of a Texan autumn; they were to be greeted by West German October rain.

An M113 passes a windmill near Nieuwams-Tordam in the Netherlands. Red and white reflective tape can be seen on the front and sides of the vehicle.

building was helped by a 1 Para platoon commander of Yugoslav parentage who had studied Russian at university and who commented, 'This guy is asking some typical para questions. It is very interesting to hear him talk and I can certainly relate to him as a fellow professional.'

Thus exercises in Europe perform manifold and apparently contradictory functions – they prepare a country fully for war while simultaneously reassuring the prospective enemy that their intentions are not offensive. Exercise planners manage to have the best of both worlds.

Armies naturally attempt to learn from experience and then incorporate the lessons into their latest training. It is true to say that control is the key to fighting a battle. A modern encounter is a fluid one, and while this may appear to suggest that no amount of training could condition someone for the experience, the practical truth is that the greater the breadth of a student's training the greater the range of his reactions and skills. Men under pressure make decisions on the basis of their past experience, be that experience theoretical or practical, so it is plain that the more

▼ Sergeant Perry, whose picture with his Stars and Stripes earned him the nickname 'Mr Reforger' after it appeared in various newspapers and publications in 1987, He was photographed at a non-tactical holding area before the exercise began.

US Air Force personnel in full NBC clothing stand guard during an exercise simulating a chemical attack on an airfield.

Anything tastes better than army food – US troops stand in line for hot coffee and doughnuts from a wagon at a 'Reforger' staging area.

US Army transport is topped up with fuel at a Dutch Army base before proceeding eastwards into Germany. The GI by the petrol pump is monitoring the vehicles to ensure that they are fully fuelled for the journey.

experience that can be drawn upon the better will be the chances of success. The Soviets are avid believers in the uses of military history for operational analysis, and such studies are obviously incorporated into exercise scenarios on both sides.

After fighting two land actions in the Falklands campaign, a senior officer in 2 Para was asked by me if there were any false conclusions that could be drawn from the campaign. 'We should not overestimate the enemy,' he replied, contrasting the badly led Argentine conscripts with the forces of the Warsaw Pact. Training for war against the Warsaw Pact is like preparing for a boxing match with a heavyweight – if you can beat him you can beat any weight.

▲ German soldiers by their Marder ICV with one of the youngsters who always gather around military vehicles on exercise. West German children are normally friendly and inquisitive, but never bad mannered. A declining birth rate in West Germany is now beginning to pose problems for the German Army.

The Falklands, Afghanistan, Central Africa, Northern Ireland, Vietnam, Algeria, Malaya and Korea have amply demonstrated that major powers have been at 'war' since 1945. Some of these wars have been against enemies of varying weight, and many have taken place in distant countries, some of which might not be perceived as an immediate threat to

Britain, the United States or Russia. Training exercises must therefore reflect the possibility of operational deployment outside Europe. In the years 1979 and 1980, for example, the British Army planned 229 exercises in 18 countries. Some 54 of these involved major units supported by 21 reserve forces, while there were 175 minor exercises supported by 63 reserve units. South America was the only continent in which British troops were not deployed. Germany had pride of place with 28 major and 97 minor exercises, but even Australia and New Zealand had their quota of minor ones.

Though many of the minor exercises can be held within the confines of a military training area, the larger-scale ones inevitably spread over into areas of civilian countryside. With questions being asked about exercise damage and cost, every army in NATO has addressed the delicate political question: 'Why are exercises necessary?' This, as noted earlier, has largely been answered to the satisfaction of domestic political opinion in Europe.

Most activity is restricted to specific training areas such as the Senne in Northern Germany or Salisbury Plain in England. Both areas (and others) have long been used for such purposes and the local population has become accustomed to such goings-on. However, the longer ranges and greater lethality of modern weapons means, as mentioned earlier, that larger areas of relatively uninhabited territory are needed than Europe can safely offer. Above small-arms level, for conventional battle group weapons, ground-attack aircraft participation and anti-tank helicopters, the vast open spaces offered by Canada provide the testing ground. The Canadian live-firing training area at Suffield has 1,000 square miles of prairie and would swallow Salisbury Plain twice over as well as the British training area at Saltau, West Germany, and still leave room to spare! Despite the acreage, only eight to ten British battle groups can be accommodated each year owing to the needs of others. NATO armies make considerable use of one-another's training areas. Castlemartin tank gunnery range in South Wales has been used by West Germany for more than twenty years, which fosters the good co-operative skills that would be needed in time of war. The demands of space and the needs of training and safety have led to the holding of support-weapons concentrations. What this

The classic image of Field Training Exercises in West Germany – a Canadian Army M113 APC parked at dawn outside Martin Frey's baker's shop. In a few hours it will be gone, leaving only a few track-marks on the road.

bäckerei v. Martin Frey
24

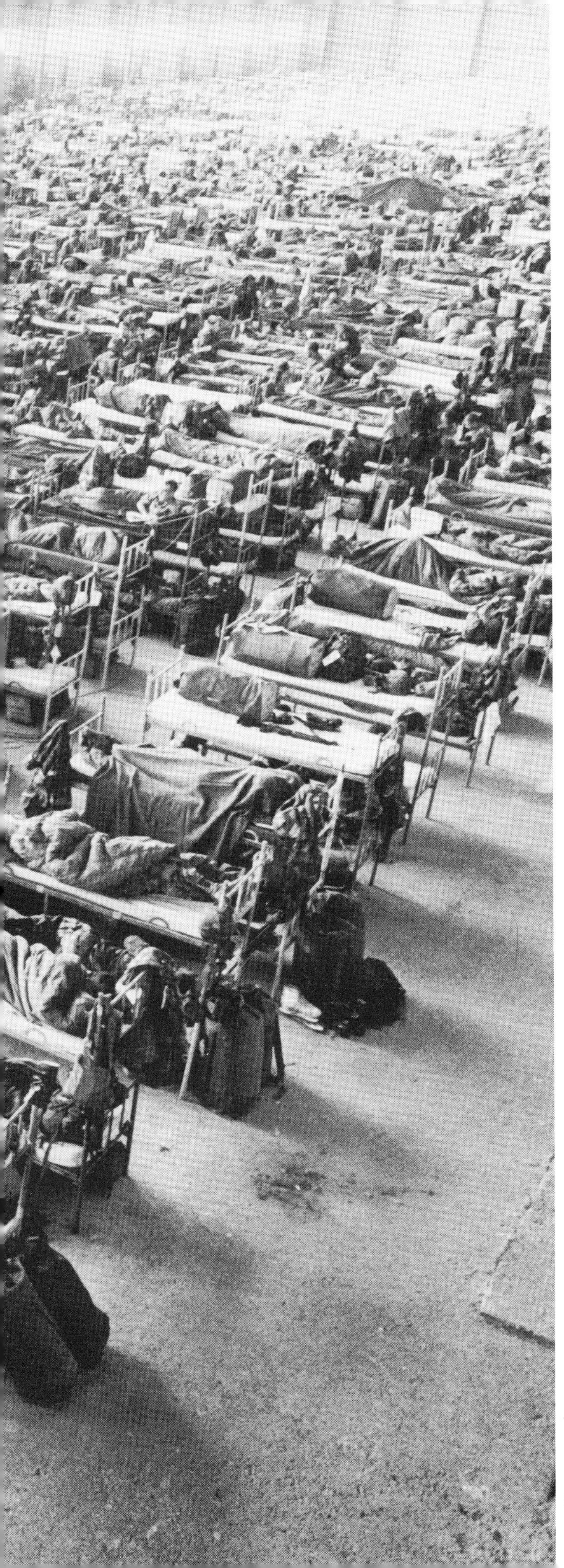

means, for example, is that the mortar platoons of the infantry battalions in a brigade or division collect at the same time and place and fire part of their allocation of live ammunition. The labour-intensive business of range safety can be shared out, and so each platoon has a chance to work as a complete unit. A competitive edge is introduced into training, but men also have the chance to meet their opposite numbers in other regiments and compare notes. At a higher level it allows new training methods and ideas to be disseminated to all the units at once.

Exercises, then, test and train the armed forces in the best and most realistic way possible in the absence of real war. They instil combat readiness in men and officers. As an experienced US officer remarked after three tours in Vietnam, 'the habits you develop in peace are the habits you take to war'. No recipe exists for success on the battlefield; the only part-truths that can be stated are that the victor is invariably the side which is less badly informed and less disorganized, and that he will therefore most likely be the better trained and the better equipped. In this context, practice makes perfect!

It may not be American ice cream, but German ice cream tastes pretty good after MREs and T Rats. Soldiers of 4th Infantry Division supplement their rations.

Bodies on the ground. Exercises can be confined to HQs, but big ones like 'Reforger/Certain Strike' need soldiers to make them work. Here US soldiers in transit occupy a huge hangar in Germany.

Preparation is a major element of a large field training exercise. It can take up to four years to plan the event, placed as it is to fit in with seasonal weather, harvests and political plans. The first visible signs that it is time for War Games in Germany is a fleet of Land Rovers busily bill-posting public warnings on trees and walls: 'Achtung manövergebiet. The locals are being reminded that in a day or two the roads will be full of grinding tracks, roaring engines and the prospect of a machine-gun position at the bottom of the garden. Veterans of BAOR and the US Army in Europe will recall the 1950s and even the late 1940s when exercises still had the character of war – almost anything went and crop damage was not the costly mistake it is today.

As a retired general remarked, 'They were still an occupied country then.' Soldiers could recall tank squadrons driving in line-abreast across cultivated fields and through hedges and fences. Anyone who has seen the damage that a tracked vehicle can do in isolation will appreciate what thirty would do to roads or fields. There were funny moments however – lack of liaison with the Germans could cause confusion. The retired general reminisced about his time as a junior officer with a commanding officer who was a hard-driving veteran of the Second World War. The regimental padre had nothing to do in the headquarters during the exercise, so the Colonel turned to him at a quiet moment and said, 'I think you ought to bury some soldiers.' It was an order. The padre organized graves to be dug, donned his stole and took up his prayer book; a small gathering of officers attended the 'service'. The padre also filled in the correct paperwork for the ficitional next of kin, and the correct markers for the 'graves'. Unfortunately some German civilians, who had watched at a distance, assumed it was a proper burial even though they had not seen coffins. The following morning they returned with several policemen and a bishop to supervise the disinterment of the British soldiers who had been 'buried' on unhallowed ground. They would not believe that it was an 'exercise' and even dug up the graves. The absence of bodies made them even more suspicious.

Today this would not happen since there would not be enough spare soldiers in a regimental HQ to dig graves; nor would there be any confusion with the locals. Bundeswehr liaison officers would be on call to

explain what was happening, and the elaborate organization that covers exercise damage would also be involved.

Overall management of an exercise rests with Exercise Control. For a medium-sized event, EC would consist of cells, each of which would be responsible for running a particular aspect. There are gunners, sappers signallers, vehicle recovery and repair, and all the other arms and units of an army that maintain the combat elements in the field. Also necessary to make the exercise 'play' run smoothly are cells for the press, civil and military police and the German liaison officer. War games on a large scale in the peacetime countryside will inevitably result in some damage to the environment, so systems must exist to ensure the military a smooth relationship with the civilian population. A crash between civil and military vehicles will have military and civil police, recovery, ambulance and press liaison teams involved. The two police forces will deal with their respective areas, and other teams will evacuate people and vehicles. The press team is normally co-located with the police in exercise control so that they can anticipate any adverse press coverage or telephone calls demanding extra information. Normally a German will handle the German press and a British or American their own nationals; in this way, confusion is avoided and the press get the story they want.

Normally there are three main types of stories written by national and local press – cost, damage and death. All exercises are expensive to stage. The cost of fuel alone for trucks and vehicles is daunting. Damage includes vehicle accidents, destruction of roads and buildings, and disruption of agriculture. Sadly, death is always a feature of exercises, though for the press the death of soldiers is of less importance than that of civilians. In Germany the weekend is sacrosanct, and so in order to reduce the risk of death on the roads during an exercise, all tracked vehicles must laager up in woods and wait until Sunday evening to move out. This is, of course, not realistic, but it does give men the chance to service their vehicles and clean up and sort out their kit. I recall one such weekend halt when the

A US Air Force A-10 Thunderbolt is serviced by its ground crew. Working in pairs A-10s are a common sight over Germany. Their ability to make slow, tight turns makes them unique in the West.

M1 Abrams moves out from a pre-exercise harbour area. The tank has a good track-to-weight ratio and causes less damage on the move, as well as being able to cross softer ground than other MBTs.

battalion had the chance to take showers at the local school. Cleaning off a week's sweat and grime was very welcome and almost a luxury.

The planning and operation of an exercise must work in time and space. So just as the 'war' must stop on Friday night, the men and vehicles must be somewhere where they can park. Units must work out some sort of training or activity to fill the two days. This is known as 'escape and evasion' and consists of interest visits to the Inner German Border (IGB).

The problems of flight logistics are immense. A vast increase in air traffic caused by fixed-wing and helicopter flights poses a very serious safety threat in a confined area. The large civil airport at Hanover must cope with scheduled civilian flights plus traffic from the Hohne training area. Thus an Allied air force liaison team must help out with flight control. It is imperative during an exercise that overflying of East Germany be prevented, since this could easily be interpreted by their defences as an invasion. Therefore, to forestall such a tragedy occurring, corridors are agreed at the planning stages of the exercise. On 'Certain Strike 87', the East German border was, for exercise purposes, moved westwards for aircraft and straightened out.

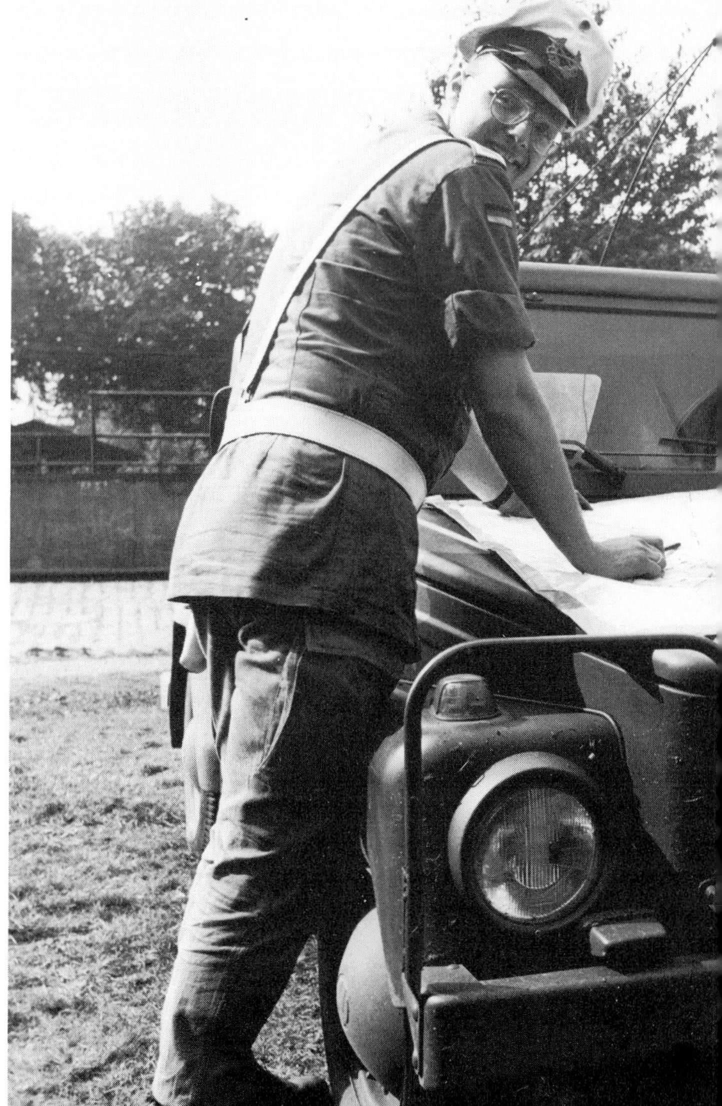

▲ On the other side of the border a 'regulator' directs traffic on a Soviet exercise. These point-duty men will show a column where it should turn off the road – maps being less widely issued in the Soviet Army. In a fast-moving exercise the country could be littered with men out of position.

◀ West German military police talk to a soldier of the 1st Cavalry as they discuss the best routes for a column of vehicles. Military police from the host nation do much to speed troop movement and reduce misunderstandings.

Flight was prohibited eastwards of this north – south line.

Night flying, power lines and cluttered air space, all make it imperative that exercise planners address potential problems well in advance. With limited air resources, Allied pilots may find that they are 'Orange' at the beginning of the exercise (and thus directing their efforts westwards) and 'Blue' in the second half, flying against targets nearer the IGB. Their home airfields may be off limits for the exercise; or, like Harrier or helicopter crews, they may be deployed in woods as part of the 'war'.

Aside from tactical air movement there is also the troop reinforcement that moves men to the exercise area. 'Reforger' – 'Return of Forces to Germany' is the exercise that tests the operability of reinforcing Europe directly from the United States. Jet lag takes on an entirely new meaning when men fly from the warmth of Texas to the damp of North Germany and then plunge into the erractic hours of a major FTX. In order to move men from even as close a location as the UK, wide-bodied aircraft are taken out of service with commercial operators and used as troop transports. It makes an odd sight to see camouflage and khaki among the pastel colours of a commercial aircraft. Again, the increase in air traffic must be planned for, and even the customs officers positioned to receive foreigners. One observer watched with disbelief as the US 82nd Airborne carried out a spectacular mass drop over Germany after flying direct from Fort Worth in Texas. As the men landed and gathered their parachutes together, German customs officers hurried around the DZ checking the rucksacks of paratroopers in case they were importing dutiable goods.

On 'Reforger' the exercise also tested the feasability of storing vehicles in Europe and drawing them from the POMCUS (pre-POsitioned Matériel Configured to Unit Sets) stores in West Germany, the Netherlands and Belgium. The vehicles are serviced in the stores and only require the installation of a battery and fuelling to put them on the road. Soldiers from the 1st Cavalry commented on the very low mileage on many vehicles. They are driven to the initial unit assembly area, where they are checked, and then move to forward locations. The total US POMCUS commitment is six divisions with their corps level headquarters support, combat service support and air support. The stores have wash racks, where vehicles are cleaned after the exercise, controlled-humidity warehouses, weapon storage facilities and ammunition bunkers.

The road move from the UK, bases in West Germany or POMCUS stores must also be pre-planned and carefully controlled. A military convoy can be a vast slow-moving jam, a gigantic roadblock or merely a huge traffic hazard. Normally routes are selected that will by-pass the cities and large towns. The routes are often given the names of playing cards, so the order may be 'Follow Route Diamond'. The soldiers should have marked the route on their voluminous maps, –

▼A West German policeman takes up a point-duty position to see US Army vehicles through an intersection. Soldiers are urged to behave correctly and relations with the police are generally good.

but, anticipating that they have failed to do so, the planners also ensure that the route is well marked out with Tac Signs. These are small road signs bearing a diamond, heart or spade according to the route, which are stuck in a grass verge or fixed to trees. Armoured vehicle-launched bridges (AVLB) may be used to protect existing bridges – they are positioned on the road, but with the front and rear on the existing bridge abutments. In this way the weight of military traffic does not fall directly on the civilian bridge. Engineers can check the class of a bridge in Germany, and thus what weight of vehicle it can take, because all bridges in Germany have an orange disk bearing the class number, which can be compared against the class of a vehicle. Thus damage resulting from overloading is avoided.

In order to reduce the danger of vehicle accidents, all NATO AFVs and many soft-skinned vehicles carry an orange hazard warning lamp. A convoy on the move by night takes on the wierd appearance of a Dantesque 'journey to another region'. It is very noisy, stinks of exhaust fumes, and is periodically lit by flashing orange lights as it rumbles by. German villagers often stand

▲ A night move on 'Certain Strike'. Here an M1 Abrams rumbles down the main street of a German town. The white and red reflective stripes on the mudguards and on the gun barrel are intended to reduce the danger of civilian cars hitting the tank in the dark.

▼ F-16A 'Fighting Falcon', the formidable American fighter aircraft. Here four of the potent machines taxi on a runway in West Germany.

Officers of 1/5 Cavalry of the 1-32 Task Force receive orders before moving out on Excercise 'Certain Strike'. They are wearing woodland-pattern battledress uitilities (BDUs) and rain-suit jackets.

A Dutch officer with his umpire's white 'neutral' arm-band watches the armour battle at Sennelager training area. Behind him are the ruins of a bunker built during the Second World War by the Germans.

'Reforger 84'. Lieutenant-Colonel William Frietas, commanding 3/77 Armored Battalion of 5th Mechanized Division, hears the bad news from the umpire group as the Blue Forces launch their attack.

Sergeant Arnold Sexton of the 241st Infantry from Fort Hood guards the road for the Blue Forces during 'Certain Strike'. He has twisted the elastic strip around his Kevlar helmet so that the luminous 'cat's eyes' do not show.

and stare; trying to conduct a normal life as a convoy passes your home town is almost impossible.

During the day children and old men come out to watch. Even the most self-confident young soldier is aware of the critical eyes of old men who served in the Second World War. One British cavalry officer remarked, 'If you catch their eye, you can see them brace up.'

So, with the men on the ground and the vehicles in place, let play commence.

Exercise play is not completely free. The story line is normally the mirror-image of the Warsaw Pact exercises that are taking place on the eastern side of the IGB. The story is in two sections – the orange forces attack and the blue withdraw and then the blue launch a counter-stroke which drives the orange back. In the 1960s the enemy were always 'red', but in an odd concession to the USSR the 'bad guys' changed from red, with its slightly political overtones, to orange. Blue remains the colour of friendly forces.

The two sides mark up their vehicles with orange or blue panels and may make other alterations to tanks to make them look like Warsaw Pact AFVs. Crude mine-rollers have been added on one exercise to make a Chieftain look like a minefield-breaching tank. In the British Army, the infantry may wear berets, combat caps or helmets to distinguish their loyalty. It is far easier when the 'enemy' is from another NATO ally, whose vehicles look very different – Marders, MIs MBTs, or Leo 2 MBTs have a very distinctive shape.

The Directing Staff (umpires) use white markings to show that they are neutral. Thay are normally equipped with similar vehicles to those used by the forces they are controlling – it is no good giving an umpire a wheeled vehicle when he is following tracks, or vice-versa. Umpires are self-contained and the umpire radio net is independent of the exercise operational nets. The problem with an FTX is the point where orange and blue meet. The logistics and road-march can be very close to the real thing; but blanks and thunderflashes cannot really imitate the effects of small-arms and artillery fire. Daniel P. Bolger in *Dragons at War* conjures up the unreality of an FTX in Germany: 'Officers and senior sergeants equipped with probability tables, radio links to "control headquarters" and a double basic load of patience, move along with participating units. These soldiers referee each "battle" using table modifiers for such items as surprise, use of artillery, visibility, and tactical deployment. This resolution results in simulated vehicle losses and troop casualties and the advance or repulse of the attacker. The control headquarters adds entertaining wild cards

2▲2-5

such as nuclear explosions, chemical strikes, air attacks, and refugees (all imitated by delays in movement and force attrition).'

'To the players,' Bolger explains, 'umpires are sources of nagging slowdowns, capricious simulated death ("you have just been struck by an enemy nerve gas barrage"), and frustrating "administrative halts", as an exasperated control headquarters attempts to restore order to the units racing around the battlefield.'

With C Company of 1/5 Cavalry, an administrative halt on 'Certain Strike' was circumvented when the men dismounted from their Bradleys and moved forward on foot. We plodded through rain-washed German countryside and head-high sweetcorn until we located a Belgian Armoured Personnel Carrier. The infantry closed with the vehicle and to their great satisfaction destroyed it. However, the Belgians still had blank ammunition and later that morning appeared and attacked the dismounted US infantry. It was a very small contact and there were no umpires present. As an outsider, I think that the Cavalry won – but the Belgians were happy to have overwhelmed the Blue forces with volleys of blank fire. 'Buttoned up' in their APC, they left the village.

Later that day the problems of control emerged again. My hosts asked me if I wanted to go on an 'air assault'. The company-sized attack would seize a bridge to the rear of orange lines and hold it to allow blue forces to pass through it before the orange forces had a chance to demolish it.

In the finest traditions of the Air Cavalry, C Company mounted their UH1Bs and were airborne. After an exhilarating treetop-high flight, the Company was into the LZ and away down a wooded path to the bridge. Part of the problem was finding the bridge – it turned out to be more of a large culvert. The near and far banks were secured, and AFVs began to appear. However, it was hard to decide if they were blue or orange. Thus far no umpire had appeared. After the soldiers had taken the bridge, and after the unknown armour (some British Challengers, some US M60s and M1s) had crossed it, an umpire did arrive. He must have been satisfied, because we were able to break for a snack supper at a Gasthaus. This made a change from Meals Ready to Eat – 'and the boys deserve it,' said the tough young captain commanding C Company.

How the umpire managed to deliver a verdict upon an attack he had not witnessed remained a mystery. Indeed, the arbitrary decisions of umpires can madden

An M-163 Vulcan Air Defence unit during a break in exercise.

An engineer marshals a unit of a Ribbon Bridge across a West German designed trackway. The Ribbon Bridge is based on the PMP, a Russian design which has been modified and improved in the West.

A combat support boat of the US Army 902nd Engineer Company is launched into the River Aller during Exercise 'Certain Strike'.

As one boat holds station midstream, a second is launched. The dawn mist is clearing on the Aller; in reality this bridging operation would take place at night. The Combat Support Boat is a British design that uses a water-jet propulsion, enabling it to move very quickly and to make tight turns and rapid halts. It is said to be almost unsinkable.

A Combat Support Boat moves downstream at speed. The fenders in the bow enable it to shunt Ribbon Bridge units around in the river, or it can tow floating bridge units into position.

With the bridge in position the first vehicles start to cross, here an M1 Abrams of 2nd Armored Division. Spare boats are on call to nose the bridge back into line if it starts to drift away from the crossing point.

The bridge is operating. In the foreground the remains of a T Ration show the pressure the engineers have been under, working through the night and early morning.

A Bradley makes the crossing. The engineer is wearing a tactical life jacket, an essential piece of equipment when working on water.

US officers from the umpire organization gather at a river crossing. White bands on their caps and arms distinguish them as umpires. They will have assessed the performance of the engineers at the bridge site.

and amaze. For British servicemen there are two exercise phrases that conjure up the unreal nature of mock war. 'What does the Pink (the paper containing the rules) say?' and 'The DS solution is . . .' The Pink is the solution to an exercise problem or the rules for judging victory or defeat. The phrase 'DS solution' is either seen as Holy Writ and the correct solution to the problem or, with more disputive soldiers, it becomes the basis for a discussion if not an argument. Exercise anecdotes abound in which either a German local or a senior British officer with experience of the Second World War dispute the DS solution on the basis of real experience in a shooting war.

The existence of Pinks and DS solutions throws into question the freedom of players to make original plans or break out of structured exercises. By their nature, exercises in populated areas are constrained by safety requirements, but every effort is made to accommodate free play. In Norway a Royal Marine major outflanked an 'enemy' position by taking his men through a long railway tunnel – the timetable had been consulted for safety's sake. As the Marines emerged to 'victory' there was a slight coolness from a senior Norwegian DS. 'Very good,' he said. 'That is what the Germans did in 1940.'

In 1980 the British 3rd Armoured Division was tasked with umpiring Exercise 'Spearpoint 80'. The Division used the normal British chain of command with umpires at each command level. In addition they were supported by nine Area Coordination Centres (ACCs) linked to an Umpire Control Centre (UCC). This structure had been adapted from a US Army system, and it proved very effective. The nine ACCs were static and had about a hundred officers and men. They were spread across the exercise area forward of the Divisional Rear Boundaries. Each ACC had four cells – Manoeuvre, Artillery, Engineer and a joint Army Air Corps/ Casualty/Air Defence cell – each of which was commanded by a major. A Damage Control cell was colocated with each ACC.

The task of each ACC was to monitor all player activities within its area. In effect, the ACCs became continuously updated memory banks for all the specialist umpires on the ground and the UCC. Umpires could update themselves by contacting an ACC. This meant that umpires had a far clearer picture of exercise play than is the case in more conventional systems where the umpire can only rely on what he can see.

ACCs also co-ordinated weapons effects, including obstacles and minefields, and ensured that each player was informed of the effects through the resident umpire in 'real time' or as close to it as possible. The

A Dutch soldier with a Redeye surface-to-air missile. His beard and long hair would be a problem when wearing a respirator or if he were to receive a head wound. This SAM is a dummy, but it has its part to play in the exercise, representing local air defence.

aim was to have a five-minute delay between cause and effect. This was achieved through the four cells in the ACC. When an artillery forward observer called for a fire mission the data was passed to the Fire Direction Centre; here the resident umpire passed details of the mission to the local ACC artillery cell. The ACC brought the local Target Area Umpire Team (TAUT) to the target, and when fire was called they simulated fire (normally with thunderflashes) and assessed its effect; then, with the Manoeuvre umpire, they awarded casualties. The results were then passed back to the ACC.

With the Engineer cell, umpires provided details of obstacle construction and breaching in their area. This was also passed on to the Manoeuvre cell. When obstacles were close to completion, Obstacle Police were positioned. When attacking forces moved into an ACC's area, engineer umpires were updated by the Engineer cell about the location and policing of obstacles. The umpires were then able to watch the breaching and repair operations and impose casualties on men and equipment. They could also impose a delay on operations if the obstacle was notional (a blown bridge or cratered road) and the engineers had brought up the equipment to cross it. Engineer umpires worked closely with the leading Combat Team Manoeuvre umpires to co-ordinate the effects of obstacles on manoeuvre play.

Casualty and Air Defence covered air attacks, the effectiveness of the defence and damage inflicted. An AD umpire would pass details to the local ACC AD cell.

Dutch engineers work on the bank seat (foundations) of a 'double single' Bailey bridge. Although an old design, the Bailey is made from man-portable loads, so that it does not need a crane to help in its assembly. It does, however, need good foundations.

The brigade HQ turns into farm outbuildings. Hessian painted to resemble walls with doors and windows can disguise vehicles very effectively against the naked eye. The advantage of locating an HQ in a farm is that civilized plumbing is often available.

The Forward Air Controller (FAC) umpire with the FAC controlling the same mission assessed the effectiveness of the strike on the target and passed this to the Casualty (CAS) cell. The two cells, which were co-located, were then able to calculate the aircraft losses or damage and the effectiveness of the attack on the target. This was then passed to the Manoeuvre umpire, who awarded casualties or imposed a delay. Details were also passed to the UCC and AD.

Helicopter attacks were monitored by an umpire travelling in the rear of the attack helicopter. Once he was in the area of the target he contacted the nearest Army Air Corps Artillery cell to find which Target Area Umpire Team was with the enemy, and they verified the target. When the attack was over the helicopter with the umpire landed by the TAUT, and they discussed the effectiveness of the mission. A joint report was sent to the ACC, and the combined TAUT and Manoeuvre umpire team awarded delay or casualties.

The Manoeuvre umpire worked with the other specialist umpires and also awarded casualties in real time, using a mix of his own professional judgement and the Armoured Vehicle Kill Potential (AKVP) system. This system is based on six principles:

1 Establish the anticipated range in which the battle has taken place.
2 Establish the kill-potential; this is based on a simple chart, which takes the range 1–1,000m as its basis.

A British brigade HQ becomes a 'brickyard'. On Exercise 'Eternal Triangle' vehicles of a brigade headquarters have been covered with hessian and canvas to blend with the stacks of bricks they are parked among.

These figures are then exchanged by the umpires.

3 The resulting figures are the total casualties that would be caused by each group in perfect conditions in every ten minutes of action at that range.

4 The figures are then adjusted to take account of target positions (e.g., in the open or dug-in) and visibility conditions. This gives the umpire the total number of casualties to be inflicted on his sub-unit during the coming engagement by his opposite number; these can be modified if good tactics are employed and proper use is made of the ground. The umpire uses his professional judgement as to when to inflict the casualties during the engagement.

5 This system obliges the umpires to build up a pattern of attrition against units that are in contact, as would be the case in war. It also forces players to take tactical decisions in battle.

6 When the battle is over, the Manoeuvre Unit notes the number of vehicles that have been destroyed and then allows the Combat Team commander to bring them back into play in order to gain full training value. Only when the combat team has one third of its vehicles operational is it taken out of the exercise for six hours to an Out of Battle Area. The time in the area can be adjusted by the Control HQ, but by dawn the following morning the unit is

Inside one of the vehicles of a US battalion TOC (Tactical Operations Center). There is the usual litter of maps, radio systems, a duplicator, acetate for map overlays, and the inevitable neglected mug of coffee.

reckoned to have received replacement vehicles and can be brought back into play.

In the post-exercise analysis the umpiring system was reckoned to have been both realistic and good training. It does however show the problem of an all-arms exercise with a wide range of weapons types. In war there would be the added factor of fear and fatigue, which would impair judgement and add to the 'fog of war'.

One factor that would not apply in war is damage control. This has become an increasingly important part of exercises in Germany. Soldiers in the Falklands were to comment that for the first time they would not have to fill in their trenches. In war the mess is left behind for battlefield clearance; in peacetime soldiers are under orders to clear up as they move. This ranges from filling in trenches and clearing up wrappings for rations and ammunition to cleaning mud off the roads. Anyone who has followed tracked vehicles after they have moved across country will know how much mud and clods they pick up. Once they reach a road the mud comes off and becomes a hazard for other motor vehicles. The rubber track-shoes are another hazard on roads, and there are tales of German drivers keeping track-shoes so that if they are in a minor traffic accident later they can claim that they skidded on the block of rubber that had come off a British/US/German AFV. In this way they can claim compensation.

Damage control is intended to minimize the cost of exercises in compensation and in terms of relations with local farmers. Soldiers have a blue card listing the cost of damage if they drive vehicles into standing

A fuel pipe snakes out to a US Black Hawk troop-carrying helicopter. Exercises also test logistics – bullets, beans and here, fuel supply.

A body on the ground. A young Grenadier guardsman cradles a belt of GPMG blank ammunition as he moves off from the German helicopters that have delivered him to an LZ. He is armed with an SA80.

An M1 Bradley parked under camouflage nets in a German wood. Modern camouflage nets are treated to give an IR signature that is similar to natural vegetation. US nets have coloured 'scrim' that allows them to be used in summer or autumn.

MRLS – Multiple Rocket Launcher System, a project that has involved most of the NATO countries, even though it originated in the United States. The MRLS will be one of the new systems to equip NATO armies.

Parachutes blossom from a C-130 during a resupply mission. By dropping them in this tight group there is less danger of excessive scattering on the ground.

crops. It has always been a source of amusement to newcomers that oil seed rape is the least costly crop – rape comes cheap on NATO exercises. The most expensive damage is to concrete roads on farms.

The Dam Con (damage control) teams possess a variety of equipment ranging from brooms and shovels to sophisticated mechanized sweepers. Stores, gravel, fence pickets, wire and corrugated iron are carried by engineer vehicles, which follow tanks in order to make good any damage. West Germany has not only a strong Green Party and anti-military groups but also a natural pride in its countryside and ecology. Tank crews are urged not to damage culverts by crossing at speed, to respect wildlife areas and to avoid spillage of fuel and oil. Telephone wire and tins can trap and maim wildlife: photographs of deer that have had to be destroyed after injury by military rubbish are used as warnings.

The Bundeswehr liaison officers attempt to minimize the social damage that may result from physical damage. I recall a German engaging me in an angry conversation as we watched Challenger tanks skid-turn off a road on to a country track. As each tank took the corner, the pavement was crushed and pulverized. As the German shouted at me I decided to smile and plead ignorance of the language. Afterwards the German-speaking British soldier with me asked me if I had understood what he said. 'I got the drift,' I replied. 'Did you realize that he said that if you were so stupid that you did not understand German why didn't you go back to your own country and smash it up instead?'

On the other hand, German farmers can do quite well out of exercise compensation. Normally crops should be in; if they are standing, tanks are urged to stick to the borders of the fields. In fact this makes for good camouflage – tank tracks diagonally across a field show up very clearly from the air. If the AFV hugs the edges it can conceal its location because the track-marks look like those made by civilian agricultrual vehicles.

There are, however, many stories of farmers emerging from their homes with a bottle of schnapps to ask a tank crew to ram the corner of an old barn or outbuilding; if the structure has been destroyed by NATO vehicles the compensation will enable him to build a new one. Equally, a poor crop may be left standing in the hope that AFVs will destroy it and compensation more than the value of the crop can be claimed. However, the liaison officers are often retired officers with good comprehension of German and a critical eye; they can see when a farmer is trying to defraud the compensation system and they can be robust in their dealings. They do support a farmer's

A US Army M-88 Recovery Vehicle. Tanks and trucks break down, often at the wrong end of muddy tracks or fields. Tracked recovery vehicles can tow them out and make running repairs to keep them on the road. The orange hazard light is almost universal for vehicles operating in Germany.

Abrams spread out as they advance. While the open ground is easy to drive over, it offers little natural cover, and the tanks would be vulnerable to ATGWs like Swatter.

Stinger in the sweetcorn. The umpiring of air attack and air defence can be very difficult, but is important because virtually all wars since 1945 have been dominated by control of the air.

claim where damage has been needless and units can be tracked down very quickly. One very sensitive area is asparagus beds – a seasonal luxury crop. Damage to asparagus has cost soldiers rank and money.

Although soldiers are ordered to tidy up and keep their equipment and weapons under tight control, there are always problems. The terrorist threat exists. Weapons, particularly compact automatic weapons, are always attractive. Generally speaking, an adult civilian hanging around troops in the field attracts attention and the men are alert. Children, on the other hand, are a constant feature of exercises in Germany. Well-behaved and enthusiastic, they ride out on their cross-country bicycles and chat in a mixture of German and English to the soldiers as they dig-in or laager-up. They can be welcome as messengers, and they will often take cash and return with food from the village shop – with the correct change! In the 1950s, as the US Army was developing the role of the Special Forces (Green Berets) in Europe, they selected German-speaking soldiers whose role would be to stay behind and organize resistance against Warsaw Pact forces. On exercise these men would attempt to infiltrate US Army positions dressed as local civilians. The most effective way was to 'borrow' a local child and appear as 'father and son'. The GIs were deceived, the local lad enjoyed it, and the Special Forces soldier completed his exercise mission.

▼ 'Quick Orders' – men of the British 5th Airborne Brigade on 'Eternal Triangle' orientate themselves and receive orders after making a landing from Puma helicopters.

Exercise mines stop men of the 2nd Armoured Division (Forward). The umpires wait by their vehicles as the 'mines' are cleared from around a bridge.

CDE Observers also cover Soviet exercises. Here Lieutenant-Colonel Harry D. Thompson (Canada) and Brigadier W. G. Bittles (UK) share a joke during an exercise in the Ukraine in August 1987.

Challenger in trouble. Even the most modern equipment can develop faults on exercise, and here an LAD (Light Aid Detachment) work on the engine of a Challenger.

Children copy adults, and on a position on the Mitteland Canal I watched a 'patrol' of 8-year-olds move with all the deliberate skill of men ten years their senior – correct spacing, watching arcs to left and right and 'armed' with mock weapons. Two had even picked up some camouflage material and attached it to themselves in the way that soliders camouflage their webbing.

All this would be fine if exercise areas were not full of hazards like tracked vehicles and training ammunition. The British thunderflash is a powerful fire-cracker, which, if exploded close to an unprotected person, can cause permanent damage. I have seen a man lose an eye as a thunderflash exploded near him, causing damage by blast and grit. Sadly, there is a NATO hand-held illuminant that looks like a thunderflash, and children have suffered grave injury when they have lit and held a thunderflash assuming that it would project a flare into the sky. A 2-inch illuminant set off in a confined space can be disastrous as the propelling rocket sends it bouncing off the walls and ceiling. Even smoke grenades can cause fires.

German children seem to be fascinated by exercises, although as they approach their teens the interest wanes since they will soon be called up as conscripts. Part of the interest has to be the novelty of soldiers in a rural community; part is the chance of being given free snacks from the army ration packs. Part, say some Germans, goes back to 1945 and the post-war attitudes to militarism and the armed forces: since the subject is still seen as sensitive, it takes on the attraction of forbidden fruit.

Teenage protest takes many forms, including slogans sprayed on walls, leaflets and posters. Very late one night, driving through a German village, the headlights of my Land Rover illuminated a teenage couple wrapped in a clinch. He did not bother to look up. The engine note and the time of night meant that we had to be a military vehicle: his right arm unwrapped from the girl to give us a two-fingered salute.

Anti-manoeuvre protests. Some are by Green organizations concerned about damage to the environment; others are by loosely left-wing groups. There is little scope for this type of protest in East Germany, although the Group of Soviet Forces in Germany (GSFG) are held in some disdain by the wealthier East Germans, who see them as intrusive.

For centuries Europe has served as a battleground for warring countries. The invasion routes that have featured during those campaigns are now well known. History often repeats itself; and in military terms this has been the case several times during this century alone – the reverse of the Schlieffen Plan, which sent German troops through the Ardennes in 1940, relied upon the

M1 Abrams of Delta-Five of 1st Cavalry Division advance across open farmland at the close of 'Certain Strike'. North Germany has considerable areas of open country, which make ideal tank going.

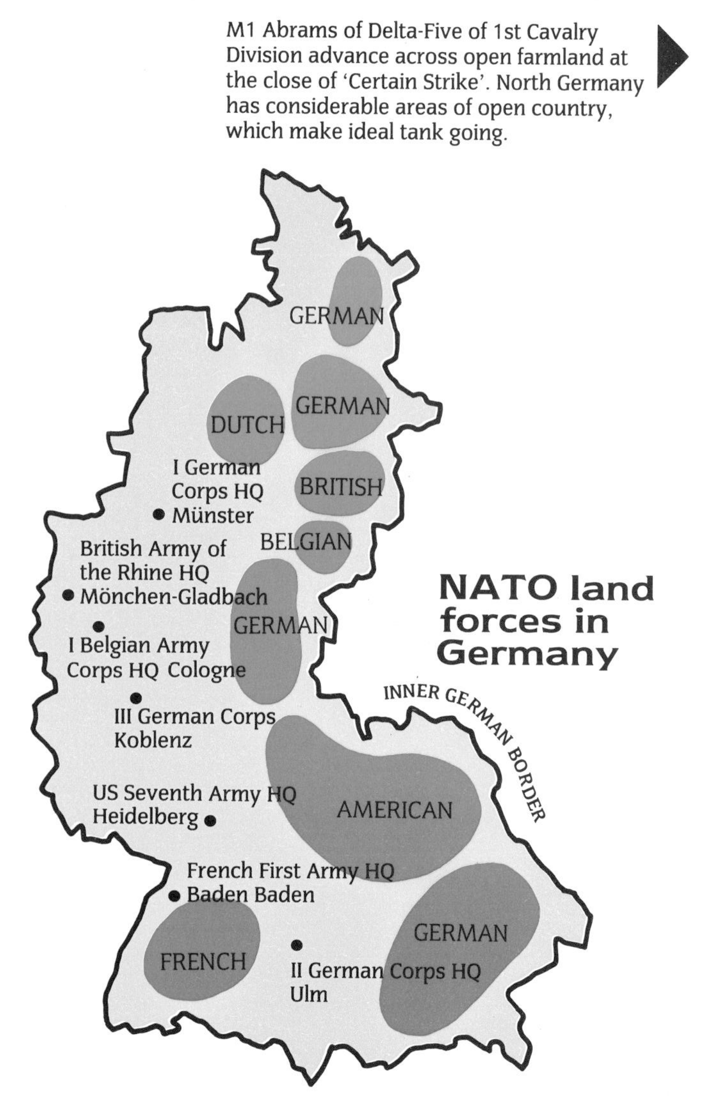

WHERE?

At the end of 'West 81' exercises, Soviet BTR60PB armored personnel carriers and T-72 MBTs parade before dismissal. Anyone who has seen massed Warsaw Pact armour on parade will testify to its impression of power.

Anglo-French forces moving into Belgium in anticipation of a replay of the First World War.

The Soviets are serious students of military history and will undoubtedly have pored keenly over such routes. Indeed, it is said that the anglers at Norwegian docks are really Russians measuring the depth of the water, and that long-distance trucks from Comecon countries always contain a driver's 'mate' who is a disguised officer checking his regiment's axis of advance. In fact an invading force would not need to go to these lengths. All the necessary information is openly available through published guide books and maps.

NATO's Exercises, sensibly enough, take place on the anticipated axes of invasion from the east, although, as we have seen, major tank and tracked vehicle movements are often restricted to designated areas. NATO's European Central Front, which covers the borders of East Germany, Czechoslovakia and Austria, is divided into two distinct zones, north and south. The US V and VII Corps, along with the German II Corps, are located in the Central Army Group (CENTAG) areas to the south, while the Dutch, British, Belgians and Germans protect the vulnerable North German Plain with Northern Army Group (NORTHAG). Each of these zones has distinct 'likely routes': the Danube Corridor leads from Austria up the river valley towards Munich and would allow the Soviets to outflank German II Corps, a benefit well worth the violation of Austria's neutrality; the Bohmer Wald area astride Bohemia in western Czechoslovakia would probably prove difficult going for a heavily mechanized army, but nevertheless it contains two potential avenues of attack in the Fulda Gap and Hof Corridor. The advantage of these would be that they sever north-south communications in NATO's defences.

It is widely believed in NATO that a Soviet assault on the Central Front would put all its weight in the

FROGs in the snow. Soviet surface-to-surface rocket crews receive a briefing during the winter exercises in Siberia.

The sight that would face the Norwegians or NATO's Allied Command Europe (Land) if the Soviet forces were to cross into northern Norway. T-72s under Captain Victor Voropayev on an exercise in the Cis-Carpathian district.

north, which is ideal tank country despite its water barriers. This would allow a rapid breakthrough to the reinforcement ports along the Channel and North Sea; it would also cut off Denmark and thereby make Norway vulnerable to attacks from the south. The disadvantage of the approach is that the Soviet left flank is then open to attack by the US V or VII Corps; but this development might be averted by Warsaw Pact forces launching spoiling attacks along the CENTAG front to engage NATO forces prior to unleashing the full weight of attack through the open North German Plain. It is to test responses to just such scenarios that exercises are held. 'Reforger', as we have seen, was undertaken to bolster NORTHAG defences, which it is expected would be needed urgently in the event of a real attack.

How likely is such a turn of events then, and how powerful are the actual forces that would undertake it? The political situation in Europe has changed a great deal since NATO was founded. It is arguable that the situation is difficult to imagine when tanks of the Group of Soviet Forces in Germany (GSFG) would cross the Inner German Border (IGB). Nevertheless a vast Soviet army has remained in Germany since 1945, posing just such a threat to the West. The Soviet presence in East Germany consists of some twenty divisions or 300,000 men divided into ten motorized rifle divisions (MRDs) and ten tank divisions, supported by an artillery division; all are Category 1 or at full wartime strength. Some 15 per cent of Soviet tank strength – 6,000 tanks – is in GSFG, while the 16th Tactical Air Army has 1,300 aircraft and 25,000 personnel to offer close support and ensure local air superiority. This quantitive strength is awesome, but the Red Army is considered to be qualitatively inferior to NATO's armies. Soviet commanders have never been encouraged to make original plans, but stick to drills, while the ordinary

538
539

solider is an often reluctant conscript rather than a volunteer. The Soviet Army was restructured to its present format following Stalin's death, and its motor rifle regiments were created to serve as a balanced force mounted in wheeled armoured personnel carriers (APCs). The MRDs were equipped with FROG (Free Rocket Over Ground) missiles in the late 1950s and SCUD surface-to-surface (SS) missiles in the early 1960s. These armaments gave GSFG tactical nuclear capability to complement their conventional weaponry. Soviet troops would probably be transported into battle in the amphibious BMP tracked mechanized infantry combat vehicle. The BMP entered service in 1967 and is armed with a 73mm gun and Sagger ATGMs, although present updating is taking place with 30mm high-velocity cannon and AT-4 Spigot ATGMs being installed. Warsaw Pact forces would be capably

Soviet marines in close-combat anti-tank drills. Tanks close-to can be very frightening, and this type of drill can instill aggressive confidence in infantry. In theory the marine will throw an anti-tank grenade at the tank and destroy it.

A Dutch officer in his Leopard IV (Improved). The tanks are from the 59th Tank Battalion, Alpha Squadron, 1st Dutch Corps.

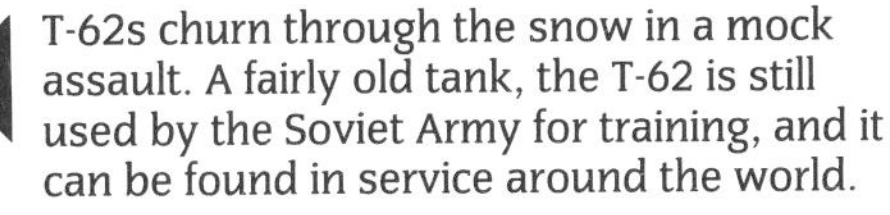

T-62s churn through the snow in a mock assault. A fairly old tank, the T-62 is still used by the Soviet Army for training, and it can be found in service around the world.

protected by fairly sophisticated mobile air defence systems, which proved their battlefield ability during the 1973 Arab-Israeli War. The USSR had supplied Egypt with ZSU-23-4 radar-controlled anti-aircraft guns and SA6 Gainful surface-to-air missiles, and these surprised Israeli pilots with their efficiency whenever low-level or high-level attacks were made against Egyptian ground forces. These systems have since been updated and improved, the SA-8 Gecko and closer-range Gaskin now being 'state of the art'. The Russians have always been keen employers of artillery, which they affectionately know as 'the God of War'. During the 'Great Patriotic War' (the Second World War) they used it in massed concentrations and still believe it to be a vital adjunct to any major attack. Until the mid-1970s, guns were towed weapons with the crew in exposed positions when deployed in the field. However, in 1973 and 1974 a 122mm and a 152mm self-propelled howitzer entered service. Their introduction gave the Soviet tank and APC forces artillery that could keep pace with them and provide supporting fire. Multiple rocket launchers, notably the BM21, a 40 tube 122mm launcher mounted on a GAZ-63 or ZIL-151 truck, are also an important part of the Soviet artillery arm. They can deliver high explosives or chemical weapons in heavy concentrations. The heavy mortar is also a favoured bombardment weapon in the Soviet armoury. Calibres start at 120mm and include 160mm and a massive breech-loading 240mm weapon.

Among the aircraft that a Soviet commander can call on are the SU-17 (Fitter-D), MIG-27 (Flogger-D) and the Hind family of attack helicopters. Weapons carried include 137mm, 190mm, and 212mm air-to-ground rockets as well as 12.7mm and 23mm machine-guns and canons.

Coloured by her experience in the Second World War, the Soviet Union has developed armed forces well equipped to fight an offensive war. The casualties (1 in 5 of the population) suffered by the USSR during 1941-5, as well as the devastation to western Russia, have become a powerful folk memory that clouds perceptions of the West. The West in turn has seen the USSR as the only country to expand its borders after 1945, taking areas of Poland, Romania, Finland and Czechoslovakia. Before 'the Great Patriotic War' began in 1941, the USSR had also swallowed up Latvia, Lithuania, Estonia and demanded areas of Finland. The countries of the Warsaw Pact (Poland, Romania, Bulgaria, Hungary, Czechoslovakia and East Germany) still look to Moscow for a lead in political and economic matters. There have been moments of rebellion – Hungary in 1956, Czechoslovakia in 1968 and a long-running opposition within Poland. Romania has its own slightly eccentric political posture, but with no common border with the West is allowed this leeway. An advance into Western Europe by the men of the Group of Soviet Forces in Germany as well as satellite forces could be mounted at very short notice. Pessimists talk of Russian troops reaching the Atlantic coast within two or three weeks. But, while the power is there, the political purpose is not. The Soviet Union uses its military power for clear political ends. Thus the Afghanistan entanglement has long ceased to make political sense: it should have been a short and overwhelming demonstration of military might, which would have confirmed the political changes the USSR had forced upon the country. Czechoslovakia in 1968 was the model. One theory holds that the USSR would launch an attack on the West if it thought that this would gain access to Western economic resources. The plant and equipment with a co-operative work force would then become part of a larger Russian and European community. However, when the USSR looks at the Second World War for its historical model it can see the amount of damage that was done to industry and how long it took factories to begin production again. War in Europe would not yield the political and economic benefits that would justify its cost. There are other options – to de-couple Europe from North America and 'Finlandize' it is far more attractive. This is a type of blackmail that would leave Europe 'free' but answerable to the USSR for 'protection'. The economic and financial power of Western Europe would flow eastwards to assist the development of the USSR. One destabilizing factor in Europe would be a unified Germany. The Russians appear to fear this most of all and have repeatedly stressed that Germany must remain partitioned. They have not forgotten Germany's power, efficiency and drive in the past. East Germany maintains one of the most respected armies in the Warsaw Pact and has deployed advisors in Angola. It is often remarked that in the event of war in Europe the East Germans would 'defect' or refuse to fight their Germanic brothers, but this argument ignores the effects of political training and the fact that old nationalistic attitudes die hard. The East Germans are Prussians; as the supreme military power among the old Germanic states, it was they who dominated until 1945. The potential enemy, then, is well armed and determined. The exercises to train the West's defences for resisting him must be as realistic as possible. The major FTX are also political gestures as much as they are rehearsals for war: hence some have obviously

Royal Marines ski-march through the snow. The ability to live in Arctic conditions stood the Royal Marines in good stead during the campaign in the Falklands in 1982.

neutral names like 'Bar Frost' or 'Purple Warrior', while others contain clear messages such as 'Display Determination', 'Certain Strike' and 'Deterrent Force'. 'Certain Strike 87' actually had the sub-title 'Deterrence in Action'. One of the FTX's functions is evidently to show potential aggressors that an attack on Western Europe would be costly in both men and equipment.

Sometimes it is hard to get away from the war and exercise role. Thus on a visit to northern Norway I saw the battalion strength Garrison of South Varanga – NATO's troops based nearest to the Soviet Union – as they deployed to resist an imaginary incursion across the border. The visit proved a valuable point about exercises. Never judge a position from the map: look at it on the ground. The road from Kirkenes is the obvious axis of advance into northern Norway (although Soviet naval infantry could make landings along the coast). The small but highly motivated Norwegian force train to block the road with demolitions and anti-armour

◀ Soldiers from Canada's Special Service Force based at CFB Petawawa, Ontario, take part in a winter exercise in northern Ontario. The SLR and Browning .30 will be replaced by the M16A1 and FN Minimi machine-gun.

▼ Towing a pulk (sled), Canadian soldiers advance in snowshoes during an exercise. North American winters are very cold but not as variable as those in Norway, and are also dry – rather than wet-cold.

ambushes. It is a road that the Germans travelled in 1945 and, as in West Germany, there are historical examples of ground that can or cannot be defended. My host showed me the rusting remains of a Soviet landing craft that had been destroyed by German gunfire – but also a valley where Soviet cavalry had surprised the fleeing bicycle-riding Germans and slaughtered them on the edges of a fjord. During our visit the exercise was in progress, and the colonel was working from a position cut from solid rock by the Germans while his soldiers occupied well-built and carefully sited bunkers that were made by men now probably old enough to be their grandfathers.

Exercises normally take place in autumn, when crops are in and damage to the countryside can be minimized. An attack at this time would encounter autumn fogs and the possibility of mud and poor going. During NATO's many exercises in West Germany, military engineers have produced 'going maps' that show the soil and drainage pattern of the terrain. Some areas are dry and firm, while others become boggy and waterlogged. If Warsaw Pact attackers were to choose the summer to launch their attack they would have the advantage of good visibility, dry countryside and the fact that prevailing winds blow westward, which would carry chemical weapons towards NATO and so reduce the chemical downwind hazard to their own troops. Exercises can be held in summer; but the problems of crops and civilian traffic make them less realistic and even less acceptable to the locals.

Familiarity with West Germany does however give the NATO defenders an advantage. How often soldiers hunched over a map in their vehicle look up and remark, 'Oh, I've been here before.' Exercises are held once every three years in an area of West Germany, so that for two years it will be left free of tanks and vehicles. However, even with this interval, the men who 'fought' over the country will remember it again. This need not be tactical, but rather a domestic familiarity – the *Gasthaus* that did not object to soldiers (some do) or the village store that had fresh vegetables and eggs for supplementing tinned composition rations. Such factors can make all the difference to temporary residence in a dank wood or hedgerow.

Tactically, local knowledge can be a major advantage. In exercise 'Brave Defender' in the United Kingdom, the Home Defence Force, which recruits from men who have had military training, either as regulars

Not winter in Canada, but winter in Germany: Canadian Leopard 1 MBTs with snow camouflage. Canada is an active member of NATO, participating in exercises in Norway and West Germany.

A Royal Marine of 3 Commando Brigade training in a Norwegian winter. He has used his ski sticks to support a 7.62mm LMG. With the move to the SA80 5.56mm system, the LMG will be replaced by the Light Support Weapon.

or reserves, but who are now in their late 30s and early 40s, was a unit without the benefits of youth but with plenty of local knowledge. The FTX was designed to test the defences of the United Kingdom against attacks and penetration by Soviet or Warsaw Pact special forces. The likely targets were 'Key Points' or KPs, which included airfields, power stations, docks and communications. Since the Home Defence Force recruits locally and deploys to local KPs, personnel had a detailed knowledge of the area. Better still, they had a knowledge of the local people, so that in war they would be able to recognize strangers if the enemy Special Forces wore civilian clothing. At the close of the exercise the UK defenders were reckoned to have won 'on points' against their attacks. However, exercises have a way of 'coming out right in the end', since the morale of the friendly forces must be sustained.

In 1980 Exercise 'Crusader' mobilized large numbers of reserves, much as did 'Lionheart' three years later, to test new ideas. Whereas men and vehicles can deploy in open fields with relative impunity, towns and certainly cities are out of bounds for exercise play. Normally, long vehicle convoys are routed around cities, and this gives the engineers a chance to build assault bridges across rivers that are normally traversed through the town. Special maps are produced of the exercise area with urban areas marked off-limits. It can even be intrusive in smaller towns – I can recall a German housewife emerging from her home in the pale light of predawn to ask if the armour and infantry battle group, of which I was part, could be quieter as her husband was ill. It was sad and at once grimly funny: how do you make a Chieftain tank engine quieter? The tanks rumbled and roared by, and it was only quiet after the last one had gone.

It was therefore interesting that 'Crusader' looked at a role for villages in a defensive grid. It is a military axiom that defence should be in depth. Mechanized

▲ Faces encrusted with frost from their condensed breath, men of the Canadian Airborne Regiment on exercise 'Lightning Strike' near Cape Dyer in Canada's Arctic.

attacks have made this even more valid. The idea that a 'thin red line' will stop an attack by massed tanks and APCs supported by self-propelled artillery is a fantasy. It is here that the German village presented itself as an ideal defensive position. Most villages are about two kilometres apart (historically they were within walking distance). However, the range of many anti-tank guided weapons is about 2,000 metres. Tactics evaluated in 'Crusader' were the local defence of villages over a wide area, the eastern side being 'hardened up' by infantry with short range anti-armour weapons while the west side had the LRATGW. Weapons like Milan and TOW would be able to hit enemy armour in the rear, where the vulnerable engine is located and also where armour is thinner. By proper mutual support, each village would be able to defend the other – so that attempts by Warsaw Pact tanks to by-pass and drive for the Rhine would bog down as they were picked off from behind. (Poor visibility with fog or rain has been cited as a problem for long-range target acquisition; improved thermal-imaging systems have proved able to penetrate even water haze, which had been a problem.)

These tactics make two assumptions: that the anti-tank infantry will dig-in and stay put, in what is a rather 'Alamo-style' stand; and that the locals will have been evacuated from the area. If this can be made to work, then villages offer the advantages of thermal screening against Warsaw Pact thermal-imaging devices – men and vehicles can be hidden in or beside buildings.

A Canadian soldier on radio stag looks out from his armoured vehicle. He has a C1 submachine-gun with a blank-firing attachment.

Vehicle crews currently carry two sets of camouflage in the British Army: netting with scrim to look like natural vegetation, and hessian sheets. The latter are lengths of crude burlap painted with the pattern of bricks or even false windows and doors. With intelligent siting and good deployment of such screens, a main battle tank can be hidden against the wall of a farm or barn.

In war the tank could be driven into buildings, and headquarters units occupy barns when on exercise. The buildings are weatherproof and offer excellent concealment – it can be a most difficult task locating a brigade HQ which has been well-sited. Land Rovers are parked in the orchard and covered in nets; HQ vehicles are inside. Movement in the open is kept to a minimum, and after dark a strict black-out is observed. Only as you approach will the sound of generators powering radios and other electronic equipment give any indications that this is the brain of a nearly-2,000-man organization. Normally on exercise the brigade major, the vital right-hand man for the brigadier, will have conducted his own reconnaissance of the area in which the exercise will take place, and he will have located farms and Gasthauses with owners willing to host the HQ for a couple of days.

This can lead to rather curious sights. At the most basic level, the HQ may be in a farmyard, and feeding and accommodation may be rudimentary. As a liaison officer for my battalion I spent many less than comfortable nights in my Land Rover curled up in a 'green maggot' sleeping bag. Around the vehicle there

was the nocturnal activity of a brigade HQ at work as well as a farmyard disturbed by these curious intruders. One experienced female journalist covering Exercise 'Lionheart' observing men in their baggy NBC suits in a German farmyard remarked that they looked like figures from the Hundred Years War on a looting expedition, intent on adding some livestock to the pot.

At the other end of the scale, the HQ can have access to a *Gasthaus*, which affords the facility of washing in hot water from a hot tap as well as the luxury of decent plumbing rather than Elsans or a latrine hole. Cooking can be done in greater comfort, and eating can be at tables indoors. In one *Gasthaus* the officers of the brigade HQ were seated dressed in their grey-green NBC suits, garments that include a hooded cowl, almost enclosing the face when it is pulled on. We were by now oblivious about our appearance, so when an American tourist looked accidentally into our room we were untroubled and explained that this was a private room. It was only afterwards that it dawned on us that the poor man must have wondered if he had blundered into the meeting of some obscure religious order seated in their monastic clothing.

Many modern German houses have cellars that can be converted to fall-out shelters by some simple alterations. They generally have good access and a robust steel frame. In war they would be the obvious location for brigade and battalion headquarters, being weatherproof and offering good protection against conventional and NBC attacks.

However, despite the examination of villages as anti-tank strongpoints and the use of farms and *Gasthauses* for HQs, the subject of urban warfare has not been addressed until recently. There are a number of reasons for this. One of the simplest is that a mock battle down the high street of a German town would be massively disruptive and unpopular with the inhabitants. They are remarkably tolerant of 'fire-fights' in the street, but employing AFVs in an urban setting would not be accepted. In a diplomatic move in Germany I removed a section corporal from the position he had chosen for his GPMG in a village we were passing through. The gunner had settled down with a good field of fire – but he had chosen the village war memorial. The dead of two world wars looked down upon him and his belted blank ammunition. I was not sure what the living would make of it.

Urban warfare can be practised in exercises in specially constructed training areas. During the Second World War the bombed streets of south London were pressed into service; but they were very British, even

◀ 'Terminator' on 'Certain Strike': a Dutch VYPR-765 with TOW under armour. This configuration allows the vehicle to fire its missile from behind cover only exposing the armoured firing post and sight.

◀ Dutch troops by their APC. Life inside an armoured personnel carrier is cramped, noisy and grubby. These conditions, as well as the fatigue of exercise, show in the soldiers' faces and appearance.

A Bundeswehr NCO talks to a US soldier on 'Certain Strike 87'. The excellent command of English by both the Dutch and the West Germans does much to smooth operations with British and American forces. ▶

very south London, and would make a poor substitute for continental streets lined with apartment blocks.

Now, however, ready-made training areas have emerged in some military camps with old married quarters. These resemble a British surburban housing estate, with gardens, attractive one-storey houses and culs-de-sac, some still with names like 'Napier', 'Wellington' or 'Blenheim'. Depending on which area you have to train in, there are acceptable levels of damage in these 'villages'. One resident near to the Longmore training area can identify the phases of an attack as it progresses through the old married quarters – the initial weight of fire giving way to the 'crump' of

Top: A Challenger tank of the Queen's Irish Hussars. Above centre: The loader rams a shell into the breech of a Leopard IV during live-fire training. Above: An M1 Abrams angles its turret.

Canadian Leopards in the summer in Germany. They have gunfire simulators fitted above the barrel of the tank. These loud pyrotechnics give an audible signal that the main armament has been 'fired'.

thunderflashes as the rooms are bombed clear and houses assaulted. A visit to Longmore clarifies why fighting in built-up areas, or military operations in urban terrain (FIBUA or MOUT, depending on whether you are British or American) can only be conducted on a training area. It is very messy, noisy and would be impossible in a real town. Tons of sandbags are packed into buildings, the roofs are shored up and inflammable objects and glass are removed. The garden is clogged with barbed wire. Sadly, even in training areas like Longmore there is a slow but steady destruction of the houses, despite training area rules and strict, if slightly unrealistic controls.

The US Army in West Germany has approached the problem with imagination and vigour and has created its own mock-German town, 'Bonnland'. This duplicates many of the types of building found in Germany. The British Army has developed good low-level urban training areas for Northern Ireland. These are for concentration on small-arms skills and observation and reaction, but have attracted foreign police and military users principally because of their sophistication. A video camera follows a patrol as it moves through the complex; and afterwards the men can be debriefed and can see on the screen their reactions when they came under 'fire'.

Orders in the field, A Dutch officer briefs an NCO on 'Certain Strike'. The combat cap is an almost universal piece of headgear in NATO, albeit with slight national differences. The zippered 'Norwegian' shirt has also found wide acceptance as a combat shirt.

British infantry carrying Clansman radios and SA80 rifles await orders to move forward. Some have taken cover in the ditch by the road, while others stand around making themselves obvious targets. Until ball ammunition or laser training systems appear, this mistake could be permanent or just embarrassing.

▲ A British GPMG gunner takes up position in a typical well-kept German front garden.

It is, however, in Berlin that urban training and exercises are treated, most seriously. Since the city would be a battlefield in the event of a European conflict, the British and US contingents have built their own training areas. The US Army has 'Doughboy City' and the British 'Ruhleben'. The French have some houses, but rely on the US Army for large-scale facilities. Interestingly, Doughboy City adjoins the perimeter 'Wall', and when major exercises take place the Soviet Army arrives and looks over the wall by means of a hovering helicopter with, presumably, cameras and video equipment. A US Army captain commented to me, 'That's OK. It shows them that we mean business.'

Doughboy and Ruhleben have buildings constructed to resemble the different types of houses, offices and commerical blocks that are encountered in Berlin. They have other features such as drains and sewers – these would be used for infiltration in street fighting. At Ruhleben the British have looked at some of the more modern buildings, and there is a multistorey block from which men practise abseiling (rappelling). The British complex also features partially demolished structures, but structurally sound, unlike real demolished buildings. These do give an idea of the urban terrain that soldiers would be obliged to work in.

Exercises in Dougboy and Ruhleben can be configured so that the battalion command post is some distance from the location in which the 'combat' is taking place. With barracks and workshops throughout the city (a relic of Berlin's wartime past), command nets can be set up and fire missions called for from guns hidden among structures. There are nevertheless restrictions on what can be used in these areas and when it can be used. Night fighting, dawn attacks on a

Tanks of the US 2nd Armored Division meet West German cows – a common experience. The alternative is a traffic jam and angry German drivers.

Sunday morning and vehicle movements are controlled A cemetery adjoins one of the areas, and a burial conducted to the sound of gunfire is an experience that the army tries to avoid.

What distinguishes training for war in a city from the vast exercises that roll over West Germany, Turkey, Denmark, Norway and Italy is that FIBUA is three-dimensional. In war in a city, soldiers can shoot downwards as well as across the terrain. Men have to learn to react very quickly, to use the high vantage that the buildings offer and to work with short ranges. In Doughboy City the US Army has built roads that are robust enough to take an M60 so that armour, either M113 APCs or tanks, can be incorporated into exercises.

The US Berlin garrison has been able to address another feature of urban warfare. Defences will not be made from dug earth and packed sandbags, but rather from the contents of houses: cabinets and closets can be packed with rubble and tables used to block doors. Here the US Army has the resources to provide the training area with time-expired furniture from its offices and married quarters. During a visit I saw two M113 APCs that had been used as road blocks. In addition to furniture and military vehicles, civilian cars are used – they can be set alight to add to the confusion and smoke that would in reality be one of the features of street fighting.

Both Doughboy City and Ruhleben have a railway system which consists of a length of track with two or three suburban railcars and a small station. Embankments make it an effective obstacle to view or fire. It is reported that the Berlin police have used this railway car facility for their own training in scenarios that assume a situtation in which an armed man has taken hostages on board a train. Police training in military urban training areas has increased as the level of armed street violence has increased. In West London, the Metropolitan Police have produced their own training area. It consists of back-lot flats as well as some three-dimensional buildings. Police train in riot clothing against 'rioters' who throw a variety of projectiles including petrol bombs.

A special area of urban training has grown up in the UK, where troops are trained for deployment to Northern Ireland. In the past, riots, snipers and mas-

German children gather around M1 Abrams of Task Force 1-32 before the tanks move out of their bivouac area.

sive demonstrations of street violence were a common feature there. The fleeting sniper in a crowded street was also a problem. Ingenious urban ranges, which feature three-dimensional targets – and also innocent passers-by – have trackways that allow the targets to cross the street and spring-loaded arms that bring the target to the window for a fleeting moment. The soldier must make an instantaneous judgement – shoot or don't shoot. There are also facilities for judging the direction of small-arms fire. In the early days of the 'Troubles', I recall a batttalion that was slated for Northern Ireland being given a demonstration of 'crack and thump', the sound of a round being fired in your direction. (The 'crack' is the sound of the round passing close to you, while the 'thump' is the noise of the cartridge exploding when the rifle is fired; the longer the interval between the 'crack' and 'thump', the greater the distance the round has travelled.) For the men due for Northern Ireland, the demonstration was carried out by the Reconnaissance platoon, who fired weapons over their heads at various ranges. It may seem rather dangerous, but it was effective. Today this type of training is undertaken with sophisticated pyrotechnics that simulate the sound of gunfire.

Urban riots, screening methods and command and control are all worked up before men move to Northern Ireland. Search techniques are practised in a group of houses especially built for this training. They have all the features of a conventional home, including resigned or bad-tempered householders and their families; these characters are acted by servicemen and women, who can be thoroughly convincing. Conviction is also a feature of rioters for the Metropolitan Police – they are normally police cadets.

Conventional military urban warfare is therefore a factor in training for war on all of NATO's fronts, but a skill that is hard to acquire except in purpose-built urban complexes. The Soviet Union has the same problems, and appears to solve them in the same way. Photographs show men moving through complexes of partially-built houses with burning tyres and pyrotechnic smoke.

In full-scale war, Soviet forces would by-pass well defended built-up areas and leave them for second-echelon troops to clear. These tactics place the emphasis on mobility and the Soviet pressure on rates of

advance. They set themselves distances that should be covered by day or night in conventional or nuclear environments. The new autobahns in Germany make movement easier and also by-pass urban areas. However, if the defenders of a city had sufficient mobility to dominate the major routes past their positions, and if demolitions channelled Soviet troops on to fewer axes of advance, such roads could become a potential killing ground. Once the lead elements of a Soviet advance had bogged down, the following units would concertina against them and become an excellent target for area weapons, either air-delivered like JP233 anti-armour weapons or by the MLRS multiple-rocket system.

Urbanization of West Germany has increased greatly since the end of the Second World War, and FIBUA or MOUT training has increased correspondingly in importance. The problem has been setting the training complexes in a larger tactical environment, which is far easier with a feature like a ridge-line or bridge in the country.

Wars are not all fought in central Europe. In fact, although the world has never been completely at peace since 1945, all the conflicts have been 'out of area'. Notable examples are Korea, Suez, Indo-China and the Falklands, all conventional or semi-conventional wars. Only the USSR has brought its armed forces into Europe – into Hungary in 1956 and into Czechoslovakia in 1968. It has also used advisors and weapons throughout Africa and become entangled in Afghanistan. So, while training for a war in Europe continues, it is also conducted in tropical, Arctic and mountainous terrain. Units such as the Allied Mobile Force (AMF) or 3 Commando Brigade go on exercise in Norway and Turkey to test their skills and familiarize themselves with climate and country. The Royal Marines send a small group annually to the Far East to train in jungle drills and skills, and the jungle of Belize provides an additional area for training. Out of area training allows officers and men to see what their allies do to combat hot or cold – thus the British Army has made an unplanned move from their shirts KF to the Norwegian zip-fronted shirt, which was orginally adopted by the Royal Marines of 3 Commando Brigade.

A change of environment is testing for equipment as well as for men: valuable lessons are learned about how oils and lubricants perform in extreme cold, the battery life of equipment and the suitability of different

◀ Given a little rain and a lot of tracks and even surfaced roads can be messed up, let alone open country.

types of rations to varying climates. In both the jungle and the Arctic there is normally a supply of water (albeit of varying purity), and this can be used to make up dehydrated rations. This theory was extended to the troops who went south to the Falklands in 1982. However, the one-man Arctic ration packs were not ideal: it was cold, but snow had not fallen, and so men were reduced to hunting about for less-than-pure water from streams. Since the 1982 campaign, ration packs now include water-purifying tablets. Troops serving in Hong Kong and training in the New Territories or Brunei have developed a jungle ration pack that includes Chinese egg noodles. These are not only nutritious but also quick to cook and very light. Dehydration powder, which was developed for children and adults dehydrated by heat-stroke, has become part of a service-man's kit along with saline drips for reviving men who are in a critical state.

Out of area training has two functions: initially the serviceman learns to live in the hostile environment and then he moves on to soldiering in it. Royal Marines who have served in Norway and whom I met in Brunei

On exercise in Canada, soldiers of the Royal Canadian Regiment cross a river. The umpires have declared the bridge 'demolished' and so this radio-operator, armed with a C1 submachine-gun, has been obliged to wade the river.

A Ferret armoured scout car. This highly mobile reconnaissance vehicle would observe the enemy and acquire intelligence in the event of hostilities.

A British Grenadier Guards radio operator watches as drill demolition charges are removed from a bridge during 'Certain Strike'.

General Galvin, SACEUR, talks to a West German major during his visit to 'Certain Strike'. Part of the deterrent effect of a major exercise is enhanced by VIP visits. SACEUR also has a chance to be seen meeting men in the field.

British infantry move along the edge of a German field. As well as reducing damage to crops, this is better tactically: cover is better, and tracks across the field are less obvious.

▲ Milan with the Mira thermal-imaging sight dug-in on 'Certain Strike'. The lack of overhead cover would make this position vulnerable to air bursts, but digging-in does give better protection against direct-fire weapons.

said that the cold was easier to live with than heat. At least with the cold there were measures you could take to get warm, while in the heat there were very few ways of escaping. Units such as the Royal Marines have also discovered that there are some men who cannot soldier in some extremes. It implies no deficiency in the man; it is just the way he is made. When this arises (and it can only be discovered through training in the area), the man has a note put in his files to the effect that he is not suited to operations in tropics or Arctic conditions. Generally, however, men will make the transition to out of area conditions with reasonable ease if they are fit. Norway can be prepared for at home in Scotland, the Italian or German Alps or rugged or cold terrain.

When a unit has a cadre of men with jungle, mountain or Arctic training they can keep standards up and self-train. Thus Royal Marines on exercise 'Curry Trail', their annual jungle training session, which culminates in an exercise against the Gurkhas in the Brunei jungle, are largely self-trained. Men who have attended jungle warfare schools teach newcomers. The exercise is conducted by a composite company from the corps, and so the skills are spread more widely.

Out of area training also tests the administration and logistics. The NATO Allied Mobile Force brings together doctors from Germany and Italy, truck drivers from the UK, helicopters from the US Army and German Army as well as men and weapons. If they were not deployed throughout NATO there would be no opportunity to test procedures and eliminate problems – in time of peace. On a grander scale, 'Reforger' not only demonstrates the US commitment to Europe but also checks that the men can be moved from the United States and that they can be united with their weapons and vehicles and moved from these depots into the field. Jet-lag and straightforward fatigue took a toll, but the vehicles worked, and with a clear period of time in the field the men were able to sustain their enthusiasm. For some it was the first time in Europe, while for others if afforded the experience of a tour with another unit. North American rain and wind is probably very like its counterpart in Germany, but there are language differences and social contrasts for the newcomer. He does not need to acclimatize too much (although the weather in the United States had been very mild and West Germany was rather dank), but he does need to acquaint himself with living in the field in another country and working with allies. 'Reforger' does this on a far larger scale than AMF (L).

WHITHER?

As has been observed elsewhere, training exercises at any level of scale falter on the final function of war: to kill or disable the the enemy and his equipment. The movement from the United States or even a battalion barracks in West Germany can be conducted with all the speed and urgency of war. The logistic system can even bring forward pallets of sand-filled ammunition boxes to simulate artillery shells or other stores. There is always an element of 'no duff' (real) casualties in an exercise, so field ambulances and unit medics can be tested with actual as well as simulated casualities. The problem arises when one soldier points his rifle at another, pulls the trigger and a blank explodes. He then asserts that the other man is 'dead', but the victim can just as forcefully argue that the round missed. A fruitless argument can ensue over how many are 'dead' after a platoon or company attack.

During one exercise a British Army warrant officer who was working as an umpire observed an attack as it stumbled slowly across open ground under a hail of blank fire. It closed with the defenders not taking any casualties and the warrant officer remarked, 'What we need is some ball ammunition.' In other words, live ammunition would have stopped an attack like this before it had advanced fifty yards.

It becomes more complicated when artillery or mortars are included in the exercise or where main battle tanks engage one another.

The development of laser systems in the mid-1960s progressed fast enough for their application to peaceful and military roles and even entertainment to be widely accepted by the late 1980s. Writing about lasers as a weapon in the spring 1988 volume of the *RUSI Journal*, Bengt Anderberg explained: 'The laser beam is characterized by its collimation, coherence, monochromaticity, speed and intensity. Collimation means that the radition emitted by most lasers is confined to a rather narrow beam which slowly diverges or fans out as the beam propagates. Since the divergence of most lasers is so small, it is usually expressed in milliradians.' Here, therefore, was something that duplicated fairly precisely the path of a rifle bullet or tank shell in flight. (In fact a laser has a trajectory flatter than that of a bullet, which is subject to wind resistance in flight as well as the pull of gravity.)

By fitting receiver sensors on the most likely target areas – head, arms, torso, turret, hull – it is possible to imitate the effect of direct fire on men and machines. The blank round triggers a pulse of laser energy which, with its flat trajectory, goes straight to the target. On the receiving end, the target may hear a near-miss signal or a hit. This is done by a small electronic receiver set normally fixed to the soldier's webbing on the small of his back. The ingenious design of these systems ensures that they cannot be switched off by their wearers: the only way the hit signal can be stopped is if the soldier lies down on his back, which depresses a switch. What this means in training is that a man or men can achieve a visible 'kill' for the first time in the history of training for war. What it also means is that a section commander can see when his bad tactics have produced high casualties. On an individual basis, a soldier realizes when he made the 'fatal' mistake that gave the enemy the chance to get a shot at him. At the close of an exercise there is the initial shock of seeing how many men have been hit and where they are lying; this can be a gap in a hedge, patch of open ground or defile. Then the 'dead' can be collected and debriefed, as can the man or men who shot them.

I have watched men who have not used systems like this before behave in ways similar to men who are in action for the first time and who have taken their

◀ Canadian soldiers with an 84mm Carl Gustav anti-tank weapon during an exercise in Germany. They are wearing laser training rigs, but the 84 does not appear to have a laser system bore-mounted.

first casualties. They were reluctant to move from cover and suddenly saw the short dash to the next bit of cover as the risky journey it would be in a 'live' situation. Afterwards they remarked that, as in life, they had seen the section corporal go down and then the second-in-command take a hit – both men being good targets because they were trying to get men to move for cover. There was a nasty moment in the section when they realized that someone, probably the senior soldier, would have to take command.

The attractive feature of these training systems is that they work when blank rounds are fired, so if the soldier has run out of ammunition he cannot make his laser training system work. Equally, the automatic weapon that can spray fire around comes into its own with the laser system: the gunner can put in bursts against bunched enemy, and they will take effect. With weapons such as the .50cal Browning machine-gun,

▲ A US Army TOW crew take position in a field of sweetcorn. The weapon has a night sight mounted above the day sights. Direct-fire weapons like TOW can be programed into the MILES system.

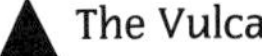

The Vulcan air defence gun system.

Dragon in ambush on 'Reforger 84'. Dragon is a reasonably easy weapon to fit into the MILES system and can be used with other training simulators.

where the cost of blank ammunition is high, the sound can be replicated from a noise-box with a flashing light. This is stowed in the ammunition box and so has the audible and visual signature of the real weapon.

The next weapon in the infantryman's arsenal to evaluate for its potential in a laser system is the hand grenade. Here a British firm, Centronics, have developed a grenade that is unique. It has not only a small pyrotechnic charge but also includes the pin and fly-off handle of the real thing. For recovery there is an audible tone that operates after the grenade has 'exploded'. Because laser energy attenuates after certain ranges, it can be adjusted to fit the spread of fragments from a grenade or the much longer range of a rifle bullet; with the laser grenade there are a series of ports around the body of the grenade, so that there is a spread of laser energy similar to the distribution of fragments. The grenade can be recovered and fitted

with a new pyrotechnic device for subsequent training. The training device is ideal in FIBUA, and has also been adopted by special units training to meet the terrorist threat.

Tracked Rapier. This sophisticated weapons system was originally destined for the Shah of Iran. It entered service with the British Army following the Islamic revolution and the UK's decision not to supply it to Iran.

Another 'weapon' from Centronics is a Claymore mine. The Directional anti-personnel mines like the Claymore are ideal for laser configuration: the fan-shaped spread of fragments from their detonation as well as their range can be duplicated. The Centronics Claymore can be fitted with a pyrotechnic charge to give a 'once only' use until it has been reactivated by the umpires.

If .50cal ammunition is expensive, 105mm ammunition for tanks is even more so. To duplicate the sound of gunfire, tanks are fitted with a bank of pyrotechnic charges that can be fired from within the vehicle. A laser system is slaved to them so that a tank makes a visual and audible signature when it engages another. It can also run out of ammunition. In order to show a hit, the tanks are fitted with an orange smoke-pot and the tank's orange hazard-warning lamp is linked to the laser receivers. If a tank is hit, the internal electrical system can also be switched off. Thus it is genuinely disabled and stops where it has been hit. Both the tank and the infantry equipment can be reactivated by the umpire with his 'gun'. The infantry cannot 'kill' a tank with the laser system on their rifles, of course. Anti-tank missile simulators using a mix of pyrotechnic and laser systems can allow infantry to engage tanks, while the main armament of the tank can neutralize the anti-tank missile position.

Even simple tasks like handing over written orders become trying in NBC clothing.

In early versions of laser training systems the information passed by the laser beam was relatively crude – 'hit' or 'miss'. Since tank gunnery depends on a variety of factors, including type of shell, angle of impact and even the prevailing weather conditions, a hit/miss system is not very realistic – in fact, it gives false information. New laser systems like those from the French company Giravions Dorand allow for the trajectory of the shell, a curved flightpath, and interrogate the firer. A tank fires and hits another; the target then asks the firer what type of shell he was using and establishes whether this would have been fatal or merely have damaged the victim. All this is done by lasers bounced between the two tanks.

Even helicopters with anti-tank guided weapons (ATGW) can be fitted with laser systems. They can attack armour and can in turn be 'hit'. They normally carry an orange smoke marker that ignites after a hit.

There are two arguments against laser systems. First, they do not allow indirect fire to be effectively simulated. Second, cover from view becomes cover from fire. This latter is tactical 'holy writ', for the thin wall or hedge that gives you cover from view is not protection against a modern rifle or machine-gun. I have watched a firepower demonstration for a Royal Green Jacket battalion which amply demonstrated this

Challenger on the move. Accurate simulation is the only way to get value for money out of systems like Challenger or Tracked Rapier.

A British company commander uses the hand set on the Clansman radio as he reports his moves to battalion HQ. Radio communications would be jammed in war, but to do this in peace is harder. However, the Royal Signals will jam and monitor nets to build up an intelligence picture and warn operators of security infractions after the exercise.

point. The young machine-gunner literally chopped down a wall with bursts of fire from his GPMG. A laser beam will not penetrate a thin layer if it is 'eye safe', as all training systems must be.

The solution to this is External Hit Internal Kill EHIK. This is a little like the systems for tanks, except that it is applied to soft-skinned vehicles. If a bodyguard squad is in training using a lightly armoured car, they know that it will stop small-arms but not anti-tank weapons. However, even soft-skinned vehicles will survive some quite heavy weights of small-arms fire and only some of the passengers will become casualties. The EHIK system fits the exterior of a building, vehicle or training range with sensors, which are linked to the harness worn by the men or women who are on the other side. Cover from view is no longer cover from fire.

Indirect fire is an interesting problem, but here Centronics and Brocks have teamed up to produce a system called 'RELACES'. This uses Brocks pyrotechnics to give the effect of artillery, mortar or chemical attack, while the laser system from Centronics make these effects more versatile. If the omnidirectional laser system developed for the grenade is incorporated, the infantry under fire can come under very convincing effects. In simple terms, RELACES consists of a laser receiver mounted on a tripod. From it run a series of cables to charges either dug-in, surface-laid or slung in branches. As the infantry advance towards the receiver, the umpire walking or driving with them can fire the charges. With a built-in delay, these explode like a mortar or artillery barrage. For anyone who has had to lay out cables across training areas, tape-off the danger areas and check the electrical continuity, this system is fast and much more realistic. The use of low-power explosives also keeps danger areas smaller, so there are none of the unconvincing lengths of white tape that are needed to demarcate the hazard areas for high explosives. In terms of realism too, the explosions happen closer to the men. Shorter cable lengths also mean that there is less problem laying out the charges and no danger of men or vehicles snagging up the cable.

AH-64 Apaches staging near Munster. They were overhauled and checked by representatives of the McDonnell Douglas Corporation.

Working on the tail rotor of an AH-64. The helicopter has been designed to make servicing easy, and the steps on the tail fin can be seen clearly.

By linking these effects with omnidirectional laser bursts, either hand grenade or two back-to-back Claymores, a pattern of 'death' or near-miss can be established. The layout can also be installed on a dug-in position so that it can come under fire and the men be obliged to keep their heads down.

Another feature of a system like RELACES is that it can be used with live ammunition. Infantry with live ammunition can advance against a position that fires back at them using small-arms simulators. The infantry can then engage electronically controlled targets that fall when hit and can themselves be 'hit' if they expose themselves to weapons firing on fixed lines – laser-triggered machine-gun fire simulators. The only problem would be that the expensive laser systems would have to be dug in behind sandbags to ensure that they were not destroyed by small-arms fire.

Historically, manoeuvres have been marked by the lack of realism that is a feature of men in the same uniform trying to look like enemies who would have different clothing, weapons and equipment. Orange and blue forces (or red and blue) have sported coloured panels, armbands or helmet-bands. Thery have modified their equipment: tanks, for example, have removed the thermal cladding from their gun barrels or their bazooka plates to look like Warsaw Pact MBTs. Variations on combat caps, helmets and berets have been worn, and combat jackets have even been turned inside-out to give a plain green appearance. The wearing of 'enemy' clothing for training has changed over the years. During the Second World War for special forces operations, as part of a pre-operation rehearsal, men might be kitted-out in enemy uniforms. The US Army in the 1950s looked at the concept of

The complex sensors and weapons load of a modern attack helicopter, like Apache, can be simulated, but the problems of servicing in the field can not. The job would be worse in the rain.

Wearing a Centronics SAWS rig, a soldier places the Centronics Claymore. This laser-operated training aid accurately replicates the fan-shaped blast of a real Claymore – except that instead of 700 steel balls it is a pulse of laser energy that triggers SAWS receivers.

'Aggressors', and Daniel P. Bolger wrote in *Dragons at War: 2–34th Infantry in the Mojave*, 'The Aggressors were marked by a green triangle in a white circle (the Circle Trigon), supposedly spoke Esperanto, and used strange weapons known as the INTERA tank and Ripsnorter antitank missile. They had Ming-the-Merciless crested helmets. It must be assumed that these Aggressors were designed to be inoffensive to real American enemies.' (If British readers are inclined to laugh, they should recall the awesome 'Fantastians', who not only had an order of battle but even a language. A training film may still exist in which the young platoon commander's voice-over is heard as he selects the men for a night patrol. 'I'll take Smith; he's a good radio operator. Corporal Brown is a good man in a tight corner . . . and Private Jones because he speaks Fantasian' – a line sure to bring the house down.)

MRLS of the 1st Cavalry Field Artillery are serviced in the field. The tracked cargo carrier chassis upon which the system is based is a rugged and versatile vehicle.

Daniel Bolger, commenting on the Aggressors says: 'Nobody really read the dull reams of Aggressor tactics and doctrine dutifully cranked out by the Army's intelligence community. Aggressors fought like Americans with crested helmets. That is, if they fought at all.' He explains that these enemy forces were pliant, outnumbered and ineffective role players. 'It smacked a lot of cowboys and Indians, with very stupid, indolent Indians.'

The war in Vietnam and the US Army's return saw the end of Aggressors (and somewhere the Fantasians slipped away too). To the US Army on exercise, the enemy now became 'Opposing Forces' – OPFOR. With time, OPFOR would almost take on the status of an élite within the US Army: on exercise with 1/5 Cavalry who had recently trained against OPFOR units I recall the respect in which they were held, 'They're good' being the honest soldier's salute to his 'enemy'.

It was the US Air Force and US Navy who first looked at training against different types of aircraft employing Soviet tactics. The Top Gun programme, which prepared pilots for Vietnam, resulted in the US Navy attaining an operational kill rate of 13 to 1 between 1969 and 1972.

A Canadian Army Military Policewoman learns how to rappel (abseil) during an exercise at CFB Gagetown. The role of women in the army has come under greater scrutiny with declining birth rates and when there is full employment.

Men of 4 Para, a TA reserve unit, leave the drop zone at Stanford in Norfolk during Exercise 'Brave Defender', a major home-defence exercise. The role of reserve units has been re-evaluated in the recent past; as volunteers, they are 'twice a citizen'.

Men of the recently formed Home Service Force guard a radar station, a 'Key Point' in 'Brave Defender'. HSF troops are normally aged between 20 and 50 and will have had previous service in the armed forces. In many ways they carry out the work that was done by the effective Home Guard in the Second World War.

The US Air Force established the 64th Fighter Weapons Squadron as Aggressors flying T-38 Talon trainers. It grew into the Red Flag programme, which trained pilots in air-to-air combat and fighter ground-attack missions against targets that had Soviet-style air defences and tactical layouts.

In the 1970s the US Army took over a large area of the Mojave Desert not far from the USAF's Nellis Base, which housed the Aggressor squadron. At Fort Irwin the Army developed their own training and tactics for OPFOR based on current Warsaw Pact practices. The OPFOR units were made part of a permanent establishment at Fort Irwin. Better still, they were armed and equipped with Soviet, or Soviet-style vehicles and equipment. By the early 1980s it had become the National Training Center (CNTC).

Any soldier will tell you that his first contact with the enemy, real or on exercise, is always rather tentative. He asks himself, 'Is that really a BMP?' or, 'Is that helmet Soviet or a NATO one?.' In these brief moments of indecision he may live or die. OPFOR were equipped with helmets made from fibreglass moulds of Soviet helmets. They had uniforms with red or black insignia according to their arm of service, infantry or armour. Though some of the detailed insignia was modified, the effect was unquestionably 'Soviet'. Coverage in the Western media reached the USSR, and *Soviet Military Review* published an article in November 1987 by Captain Vyacheslav Sedov entitled 'Camouflage Games: The way anti-Soviet psychosis is whipped up in the US armed forces'. He wrote: 'The Army programme for a "real facing enemy" rests firmly on the official directive AR 53-2 which, as stressed in the *Armed Forces* magazine, makes it expedient to establish special formations simulating the enemy. Thus in the national center for US Army combat training at Fort Irwin (California) there appeared a "32nd Guards Motorized Infantry Regiment" (!) numbering 1200 "Russians". The land forces personnel training program with the ominous name "Red Thrust" presupposes that instructors of the "enemy simulation subunits" acquaint American servicemen with Soviet Armed Forces' fighting methods. The do give it hot to the "32nd Guards Motorized Infantry Regiment". Small wonder: it has to fight all units of the US Army.'

The men who make OPFOR work at the National Training Center, Fort Irwin, are tank crews from 1st Battalion, 73rd Armor, and infantrymen from 6th Battalion (Mechanized), 31st Infantry. They are permanently stationed at the NTC and are armed and equipped to resemble a full-strength, first-line Soviet regiment. They do in fact field a certain amount of genuine Soviet equipment captured in the Middle East and traded by the Israelis. It can be hard to keep this running as spares have to be found, or reverse-engineered, or new engines fitted. An easier solution has been to take M551 Sheridan Armored Reconnaissance Airborne Assault Vehicles and modify them with false turrets and plates to resemble T-72 tanks, BMP-1s, and ZSU-23-4 and SAU122 SP guns. Sand and green camouflage and large call-sign numbers assist the illusion. The BRMD-2s were created from M880 trucks with the addition of missile, chemical or air-defence equipment.

OPFOR has the advantage that it has trained over the National Training Center since late 1981 and so knows its terrain very well. (In real life, an attacking Soviet force would be penetrating unknown territory in West Germany or Austria.) The scales are further weighted in OPFOR's favour: all weapons systems and vehicles are assumed to work faultlessly. The defects in the BMP – such as its tendency to catch fire – do not materialize with the M551 version. And behind the Forward Line of Troops (FLOT) the 'Soviet' logistic system functions better than it does in reality. The final advantage that OPFOR enjoys is much greater freedom of manoeuvre and flexibility of command. Soviet commanders work to drills and strict orders – good in the steppe, but not in congested western Europe or the mountains of Afghanistan. Daniel Bolger aptly describes OPFOR as being 'the Russians as they wish they were'.

What makes the NTC so effective as a training tool – and points to the future – is the extensive use of laser-based range instrumentation. Individual riflemen as well as armoured vehicles are linked into the Multiple Integrated Laser Engagement System (MILES). Hits are recorded, and automated communications remove the 'fog of war' for the central umpire unit, while commanders on the ground have only the information that call-signs are off the air or units destroyed.

The training package at the NTC falls into two broad areas: Force on Force training against OPFOR and its vehicles with MILES supported by Observer Controllers (OCs); and Live Firing Training (LFT) against computer-controlled mechanical targets. (Anyone who has made the switch from blank to ball, be it 7.62mm or 5.56mm to 105mm or 120mm, will recall the shock of this.)

The Observer Controllers are a vital part of the 'war' in the NTC because they can assess the effects of indirect fire weapons such as artillery and mortars and fighter ground-attack aircraft. Armed with a controller

A Soviet T-54 MBT used by OPFOR. Much of the equipment has come from the Middle East, where the Israelis have traded captured equipment with the US Government.

OPFOR with an M-551 Sheridan modified to look like a Soviet ASU Airborne Assault Gun. Gunfire simulators are visible above the barrel, and call-sign numbers have been painted on the turret to imitate the Soviet practice.

A Dutch Leopard 2 MBT during 'Eternal Triangle'. Some weapons are 'force multipliers' because they are powerful enough to compensate for their relatively small numbers: the Chobham armour and 120mm gun of the Leo 2 make it such a 'force multiplier'.

OPFOR cut a dash on a Sheridan modified as a BMD. The drab uniforms and black berets that distinguish the 'enemy' can be seen, as well as Soviet small-arms (AKM) and an anti-tank weapon (RPG-7).

Tracked Rapier, another force multiplier, with its combination of mobility and speed into action with protection for the crew.

gun, they can 'kill' and 'revive' men and vehicles and add CS gas to enhance the realism of an NBC attack. The OCs make sure that the proper man or vehicle evacuation operations take place before they revive them; thus a unit can take percentage casualties and really feel them as combat effectiveness declines.

Since the NTC is a vast unpopulated area the live-firing exercise can be conducted on a grand scale. At the NTC the only non-live ammunition are the anti-tank guided weapons TOW, Dragon and Viper. These are MILES systems triggering sensors in the targets. However, the targets are fitted with black smoke-pots, so that a hit produces convincing results.

The live-firing phase consists of a day defence, a night defence and a 30-kilometer advance to contact across obstacle belts and the rugged desert terrain of the Mojave.

The defence shoots are like a gigantic version of a British ETR (Electric Target Range). Although the flip-up targets are in fact static, by ingenious progaming they appear as men or vehicles that move from cover in short dashes towards the defended position. The targets even 'fire' with pyrotechnic simulators. If a 'vehicle' is hit, it will remain stopped, as in real life. As the BMPs and BTR60PBs reach an assault position they 'stop' and clusters of 'infantry' appear in the final rush. For the controlling staff the defence exercise is relatively easy: they can observe the shooting and stop it if there are safety problems. Advance is another thing. At the NTC the enemy are in platoon and company positions, surrounded by real wire and anti-tank ditches. The mines in the anti-tank minefields are dummies. Combat engineers can use explosives to breach the minefields, and the participants are under pressure to get through them and on to the next position. The OCs are present, 'killing' men and vehicles and assessing performance.

Daniel Bolger recalled that his unit did four days FFT, four days LFT, then a final five days of force on force. The armoured unit that operated with them had a nine-day FFT and ended with a live-fire exercise. The Observer Controllers were constantly present, 'to teach and to watch, and they taught through the use of the After Action Review (AAR). The senior controller was the chief, operations group. The COG was a craggy, hulking full colonel with a voice like shifting gravel and a mind like a steel bear trap.' (Good qualifications for any umpire or directing staff, since players are disinclined to argue.) OCs covered the performance of a unit down to platoon level, and at company level they based their teaching and evaluation on The Seven Operating Systems:

1. **Command and Control**
- Troop-leading procedures
- Facilities (TOC, alternate, command group)
- Communications/Electronics

2. **Manoeuvre**
- See the battlefield
- Fight as a combined-arms team
- Concentration of combat power
- Use the defenders' advantages
- NBC defence

3 **Fire Support**
- Tactical Support
- Artillery
- Mortars

4. **Intelligence**
- Direct collection
- Collect information
- Process information
- Disseminate and use

5. **Air Defence**
- Support scheme of manoeuvre
- Employment

6. **Mobility/Countermobility**
- Mobility (breach obstacles)
- Countermobility (build obstacles)

7. **Combat Service Support (CSS)**
- Plans and facilities
- Vehicle recovery
- Maintenance
- Supply
- Administration
- Medical.

This procedure for debriefing has been developed at the US Army Combined Arms Training and Doctrine Agency (CATRADA), which is affiliated to the Command and General Staff College, Fort Leavenworth. The value of the Seven Systems is that students and OCs have a clear structure within which to work. Nothing can be overlooked or overemphasized.

At battalion level an AAR took place about two hours after a contact was over and lasted between two and three hours. Bolger continues: 'Embarrassing questions and comments were the rule, and the senior controller explained doctrinal errors as he went along.' For the commanding officer of a battalion there were other observers sitting-in on the AAR, and at the end he could expect comments from the COG, the Fort Irwin commanding general and visiting assistant divisional commanders. The colonel of 1/5 Cavalry, a veteran of the 82nd Airborne in Vietnam, recalled his experience of the NTC as we waited in a lush German village: 'It

▲ Challenger. Its armour and mobility are good, but the fire-control system needs to be improved if results from the Canada Cup NATO tank competition are a reliable guide.

▼ 'It is, after all, their country' – a West German Marder MICV camouflaged with live vegetation. Ecological pressures now prevent most NATO armies from using this.

was as hard as war.' Which bears out the old adage, 'train hard – fight easy'.

Given its reputation, the NTC is a useful goal for units to train for. The 2nd Battalion (Mechanized), 34th Infantry, worked up through a series of exercises to prepare for the challenge. These included company exchanges with the 2nd Battalion, The Royal Canadian Regiment, and a Combined Arms Tactical training Simulation at Fort Leavenworth. It was on these exercises that staffs and sub-units would shape up and sort out problems.

'The battle' can be recorded. In After Action Reviews, success and failure can be analysed and displayed using slides and diagrams. Units can be given tapes of the action for their own self-instruction after they have returned home. Perhaps one of the defects of the US Army system is not that they do not train sufficiently hard and realistically – the NTC and subsequent deployments to Europe show a high level of commitment – the problem lies in the individual postings system within the Army, which does not allow a unit to build up a cadre of experienced commanders at all levels. Daniel Bolger comments that within days of their return, The Dragons, 2–34th Infantry, were already losing men to other posts in the Army and so were losing the training value of 'war' against OPFOR.

Fort Irwin and the National Training Center with its dedicated 'enemy' is a rich army's training solution. It is also probably one of the best. With less technological expertise and space to utilize, Battle Group Trainers were the solution arrived at by NATO's different European national armies in ways that suited their doctrines and techniques. In the British Army a Battle Group Trainer generally consists of three areas: a computer room that tapes 'radio' conversations and

A French soldier armed with his 5.56mm FAMAS. The French are politically still part of NATO and are taking an increased military interest.

Street fighting men. TA paratroopers prepare to ambush Dutch Leopard 2s in an ideal enfilade position. The increased urbanization of West Germany has forced planners to look at street fighting more seriously.

▲ Sheridans modified to look like T-72s take cover in an anti-tank ditch, as OPFOR forces prepare an ambush.

creates the noises of jamming or signal attenuation; a headquarters area where a battalion or battle group is set up; and a large room with a map-table and markers. Around this map-table are grouped the company commanders and representatives of the specialist arms attached to the battle group: gunners, engineers, reconnaissance troops and armour.

The Battle Group Trainer is a useful test for the colonel, who occupies the headquarters area, and for his immediate HQ staff, who are in the area. Normally the room is fitted out to look like an HQ in the field. The HQ may be in a barn or a wood. Vehicles are clustered under camouflage nets with 'radios' bolted in position. In reality the vehicles are un-roadworthy wrecks carefully repainted; the radios are sophisticated simulations equipment; and the wood or barn is a careful creation of timber, fake grass and painted hardboard. But once the exercise 'play' has begun the illusion becomes overwhelming, just as modified Sheridans become BMPs and T-72s in the NTC.

An exercise can take place over two days. The first day is spent on a Tactical Exercise Without Troops (TEWT), when the headquarters elements and sub-unit commanders walk the ground. (One of the good features of a Battle Group Trainer is that it is related to real ground, normally in the near vicinity.) The colonel will site his positions and the company commanders will site their weapons. Then, at the end of the day, there is a chance to discuss the defence and possible enemy courses of action.

A Warrior MCV aims its 30mm Rarden gun to cover the men who have deployed from the rear doors. The turret-mounted gun will give British Infantry their first troop-carrying vehicle able to provide them with intimate support. ▶

On the second day the HQ retire to their 'wood' or 'barn' where, clad in NBC protective clothing and even working in respirators, they monitor the war and pass orders to their company commanders. The company commanders are in the comfort of the map room in shirtsleeves or sweaters watching 'movers' in their gym shoes carefully moving across the map board placing symbols to show the enemy attack. As the players see the enemy approach their position they pass reports to the HQ. Some soldiers are conservative souls who cannot enter the drama; others raise the

161

temperature, introducing fear and tension into their voices as the attack develops.

The training value of a Battle Group is considerable. One of the chief attractions is that it is cheap to run, although the initial cost is considerable. All 'radio' traffic is hard-wired within the building so that no-one can eavesdrop, and it can also incorporate ECM as well as the facility to tape and debrief. Video is proposed, which will allow the map board to be examined while the tapes run. Under stress, men will commit errors of procedure or security that they will not remember after the exercise is over. Taping is therefore essential.

The Trainer is principally intended to test the HQ, but even at platoon level officers can see how their dispositions have worked. A computer score system makes allowances for range, angle and weapon type as tanks are engaged in the contacts. If the 'enemy' has a good grasp of Soviet tactics he can give both players and the HQ indications of where he intends his main attack to fall and where chemical or artillery attacks have prepared the way; it is up to the intelligence officer in the HQ to evaulate the information and brief the colonel.

The enemy will have given the visiting unit no guidance on how he intends to attack, and so there is training value also for the men in the map room.

Distinctions have to be made between 'map games' and 'war games.' In effect the Battle Group Trainer is a map game. It tests procedures and the team drills of an HQ. It does not test tactics, which is a secondary feature, because the information on weapons performance is not available. 'War games' in their most precise sense are more highly classified and test tactics and techniques using classified information and computer analysis. A map game in a Battle Group Trainer is in 'real time'; the battalion stands and fights for a day. A 'war game' can take a tank-versus-tank contact – which is over in minutes – and after it

◀ MRLS crews under camouflage netting share a smoke. The MRLS has the capacity to swamp its target with heavy fire and will also be able to deliver terminally guided sub-munitions to targets like massed tanks.

It may have been a successful film and TV series, but it was based on the reality of military medicine. The 16th MASH, which flew out from Fort Riley, Kansas. ▼

An M-551 Sheridan modified to look like a BMD at the National Training Center, Fort Irwin. The flag signal, typical of Soviet tactics, allows vehicles to move in radio silence. The MILES harness can be seen on the soldier on the motorcycle.

RELACES showing the umpire gun being 'fired' at the laser sensor-head, which in turn has passed the signal to the smoke-pot. Sophisticated pyrotechnic battlefield effects can be created using the system.

OPFOR soldiers debus from an MTLB. This is an authentic Soviet vehicle. Spares and servicing for these veterans of Middle East wars can be a problem.

A gunner takes aim behind his SAW (Squad Automatic Weapon), an FN-designed light machine-gun that fires 5.56mm rounds so is compatible with the M16A1. A blank-firing attachment has been fixed to the muzzle.

has looked at the variables and conflicting information it may not have reached a clear decision at the end of the day. It is strictly scientific and analytical.

In fact, the Battle Group Trainer, is a very sophisticated version of the old telephone exercise, where field telephones were linked within a small circuit, and players occupied rooms or buildings where they monitored maps. The attraction of the Battle Group Trainer, like the NTC, is that the umpires are outside the exercise and can monitor it impartially. A telephone battle organized within a unit is hard to administer – the second-in-command is normally in the 'hot seat' while the colonel is chief umpire.

The applications and uses for technology in the military situation are endless. Already developments such as the high technology exercise and training ground have been transferred to countries other than the major powers; the Jordanians have a simulator-based system for infantry and armour, the Chinese People's Liberation Army has a system, while Chartered Industries of Singapore are producing one that will doubtless be sold to many countries. As the cost of the equipment comes down and the elements of it become more sophisticated, so more exercises will be fought in their entirety using just high technology. This precision will force soldiers to take greater care with their fieldcraft and camouflage and thus improve things where they are vital – the quality of the individual fighting man. The problems of logistics, and command and control will still exist, as will the need for quick and accurate communications to keep players and controllers informed of developments and 'scores'. But the advances promise to improve fieldcraft and military readiness and efficiency beyond everybody's expectations.

A shelter half on the side of a Stalwart fuel carrier provides 'home' for British troops in a non-tactical harbour area.

Living the the field is as much part of the test of an FTX as the tactics of the 'war games'. The ability of a unit or an individual to administer him/themselves away from home or base is vital in war and so is tested in peace. In the Arctic, bad administration means frostbite or death; in the jungle men can die from heat exhaustion; and in the desert there is dehydration as well as heat-stroke.

Some of the skills of living in the field will have been taught to the soldier during his basic training. However, practical experience is a very good teacher. I can recall the gloom as soldiers dug into their rucksacks to try to find some small 'home comfort' only to find that they had not packed it. Equally men have carried far too much on exercise. A veteran of the Falklands said that he only needed two pairs of socks – one on his feet and the other padding out his rucksack straps where they dried against his shoulders. The British Army is introducing new load-carrying equipment, but the concept of living out of your rucksack, fighting from your webbing and surviving from the contents of your pockets is still valid. Men do become lost on exercises. An ambushed night patrol can send men in all directions, and I can recall a soldier who was missing for a night and most of a morning after his patrol was 'bumped'. It is essential, therefore, that men carry the right equipment for the job. Visitors to exercises may not experience the drama of a night ambush, but in the cold and wet of North Germany they will wish that they had chosen the right clothes for the job.

An exercise is finite, and this makes surviving a great deal easier – even in the grimmest weather the soldier need only count the days off. The regular will be happy to settle down to as quiet a routine as possible, keep dry, cook his food and count the days. The volunteer reservist is full of enthusiasm for the novelty and wants to have as much activity in his two weeks as can be crammed into fourteen days. He has only the routine of his civilian job to look forward to when he returns to the UK. What both groups have in common is the country and the weather.

Life in the field can produce a philosophical approach to life in general. The lunacy of adminstrative halts, the long delays as transport fails to arrive or when a flight is unavailable, the 'helicopters' that turn out to be 4-ton trucks, all produce a pessimistic acceptance of life's deficiences – especially if it is raining.

There is an enormous difference between the exercise on the local training area and the same in, for example, Brunei or Oman. In Exercise 'Curry Trail', the annual Royal Marine jungle training deployment to Brunei, many men in 1983 participated who had not been on Operation 'Corporate' in the Falklands in 1982. It was their compensation for staying in the UK and keeping the headquarters and supply units operating while others went to war. Many of the US soldiers on the Reforger exercise had served in West Germany before; they were, however, as interested and enthusiastic about the visit as first timers.

One of the ironies of attending an exercise in any capacity, other than as a soldier, is that there are no 'correct' clothes. The soldier wears combat kit, which is universally acceptable; the visitor, be he journalist, politician or representative of a defence manufacturer, can end up with variations on the theme of gumboots and a business suit. Every job has the right and wrong clothes and useful kit and equipment. If a civilian visits a unit on exercise he will certainly be kept warm and dry, for NATO armies are well aware of the value of good publicity. Neglecting a visitor is the best way to get bad PR. The Royal Marines assume that their visitors will not come equipped for either the extremes of the tropics or the Arctic. Thus the first visit is to the unit clothing store. Here the visitor will be kitted out. The Royal Marines, assume that if you are prepared to endure the climate of Norway or Brunei then you also want to get into the field. This experience is well worthwhile and gives the visitor an idea of what 30°C below really feels like – very cold.

Being dressed in uniform has advantages and disadvantages. The clothing is sound and generally weatherproof. However, if the wearer looks wrong in it, normally because buttons or zips are left unfastened, or boot-laces trail, he makes himself an object of unspoken ridicule. The ideal happy medium with sustained visits to units in the field is to come as a competent back-packer would approach a walking holiday. Very colourful outdoor clothes are not popular with men who are trying to conceal themselves or their equipment, but sound footwear, a weatherproof jacket and a sleeping-bag are essential.

The service PR team may supply a vehicle and an escort and driver, which makes the visitor mobile and reasonably independent but can attract attention because of the numbers involved. The advantage of having the freedom to be able to stop or move when you want, rather then when the battle group or combat team decides, is very welcome. A good escort officer and driver are to be valued; a poor escort can cramp your style. However, if you are fortunate, you may be left alone with a unit, with no escort at all. I was very lucky to be able to visit the US 1/5 Cavalry on Exercise 'Certain Strike'. After we had both had a look at one-another and I had spent a couple of days and nights with them, we were on easy terms. A congenial 'outsider' can bring some welcome interest to an exercise. A female journalist of my acquaintance spent time with the 5th Royal Inniskilling Dragoon Guards on

▼ 'Hot coffee?' – US soldiers collect their evening meal. T Rations with coffee, skimmed milk, peanut butter, jelly, bread and two MREs for breakfast and lunch. To the British palate, it seems very good – but then the US soldiers were eating Compo and loving it.

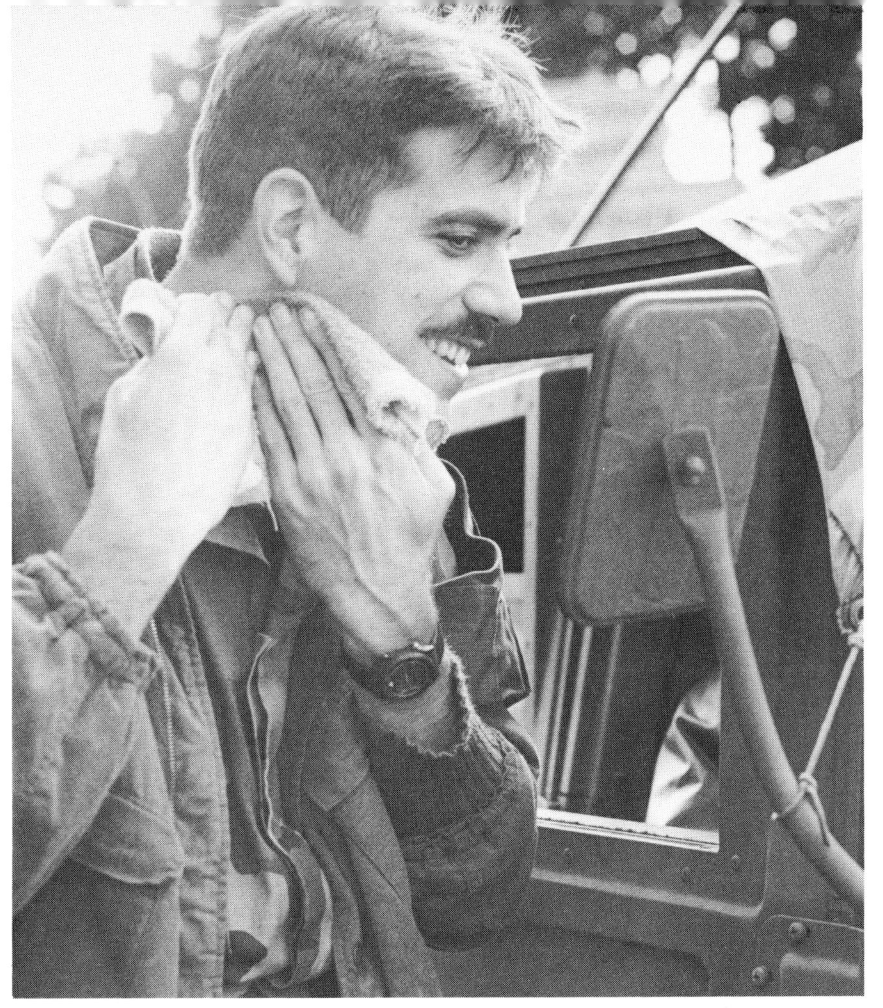

A US soldier uses a vehicle mirror to shave. Life in the field has been described as doing the private things of life publicly. It takes experience and patience to live comfortably on exercise, but it can be done.

A real chow line as American troops collect supper from a mobile kitchen unit. The trays are cardboard and the spoon and fork are airline-style plastic throwaways. There was no refuse visible, despite this profligate style.

An unlikely but timely arrival, as US troops collect their MREs – Meals Ready to Eat. German mobile ice cream, bratwurst and grocery vans have a way of locating soldiers in the field and provide a welcome supplement to issue rations. There is also a lively trade in swapping rations – US rations take on an exotic appeal to, say, British servicemen.

T Rations at the end of the day. Even if rations are the subject of complaint, they are normally well balanced and nutritious.

Exercise 'Lionheart' in 1984. Her ability to look after herself, as well as a good sense of humour, clearly made her a welcome presence in and around the command post and with squadrons in the field.

Over the years both on exercise as a reserve soldier and as a journalist I have built up a stock of equipment that has proved of value. As a soldier, it is useful to be able to make running repairs to oneself or one's equipment; as a journalist it is a quiet way of showing that you can look after yourself. Anyone who travels will have his or her own priorities and will probably add and delete from my lists.

Field Training Exercise Survival Kit

- ○ Torch with spare bulb, red filter and batteries.
- ○ Sewing kit (from a hotel).
- ○ Boot cleaning kit, spare laces.
- ○ Lightweight rucksack and larger grip with carrying strap, or larger rucksack with smaller grip. Grip takes 'civilian' needs and second-line equipment.
- ○ Lightweight sleeping-bag.
- ○ Thermal liner for bag (can be used in lieu of bag).
- ○ Goretex bivvy bag (sleeping-bag cover) should be large enough to take rucksack or kit as well as sleeping-bag.
- ○ Knife, fork, spoon, can-opener, bottle-opener/corkscrew, pint mug (which can double as mess tin).
- ○ Washing kit, containing unbreakable mirror, battery razor, tooth-polish in tough plastic container, no aerosols (RAF cargo regulations) soap in tube, wind/sunscreen cream (unperfumed), folding travel toothbrush, cucumber-scented travel wipes (a quick, water-free way of cleaning face and hands, which can also be used as lavatory paper), box of lavatory paper or handbag-size tissues.
- ○ Polythene bags to take notebooks and wallet on person. Larger polythene bags for clothing in rucksack and to take dirty clothing.
- ○ Polythene food-boxes for stowage of kit in rucksack (waterproof and will protect fragile equipment).
- ○ Portable alarm-clock.
- ○ Masking or gaffer's tape (black or olive green).
- ○ Swiss Army pocket knife.
- ○ Leatherman's tool (folding pliers and blades and screwdrivers).
- ○ Nylon 'parachute' cord.
- ○ Passport-size photographs.
- ○ First Aid Kit: painkillers, indigestion tablets, diarrhoea tablets, mixed band-aids including butterfly plasters and plasters that can cover both small and large cuts; tweezers, insect-bite cream, insect-repellent, lip salve, nurse's scissors, first field or wound dressing (x 2), toothache kit (oil of cloves), blister kit (moleskin, plasters, padding, etc.), Tubigrip elastic bandage for ankle, knee or wrist; dry dressings for burns and plaster to secure them.
- ○ Cotton money belt.
- ○ Phrase book for the land where the exercise is being held.
- ○ Commercially purchased maps of the exercise area.
- ○ International credit cards and £30-£50 (or about $50-$100) in local currency.
- ○ Waterproof notebook and pencil.

Clothing for civilian visitor to the field

- ○ Neutral (brown, grey, green, beige) waterproof jacket (Goretex).
- ○ Boots (black or brown, high lacing) Thinsulate-lined.
- ○ Shoes (black or brown, well made).
- ○ Spare trousers (smart enough to be worn out in the evening).
- ○ Spare sweater (possible evening wear).
- ○ Shirt, for evening use or field, plus tie.
- ○ Gaiters (Goretex).
- ○ Overtrousers (Goretex with long side-zips).
- ○ Boot-liners (Goretex).
- ○ Silk square of scarf as bandana (a dash of quiet colour?).
- ○ Leather gloves with silk liners.
- ○ Wool/cotton working shirt for field, with button pockets.
- ○ Silk underwear for colder weather (since silk does not have the fire-hazard problems of man-made fibres).
- ○ Trousers in drill, gabardine or wool, loose-fitting; do not wear tight jeans.
- ○ Socks, 3 pairs wool thermal, with padded soles.
- ○ Socks: thermal inserts.
- ○ Quilted liner with zip-front (and zip-fastened pockets if possible).
- ○ Hat, neutral, oiled cotton 'bush hat' style.
- ○ Trouser blouser elastics.
- ○ Underpants, two pairs (one on, one off).
- ○ T-shirt.
- ○ Multi-pocket vest (photographer's/fisherman's) or modified outdoor jacket with spare pockets.

▲ Another sort of survival – battlefield armour, here Challenger. New armours generate new anti-tank weapons, and Soviet solid shot is reported as being able to penetrate ceramic 'Chobham' armour.

For soldiers there are, of course, restrictions on dress. This is not merely a form of discipline; it is sound tactical sense. In darkness or half-light the shape of a uniform, helmet, webbing and combat jacket are often easier to identify than the individual. If men wear informal equipment in the field it is harder to recognize them. Another good reason for using 'issue' equipment is that spares and replacements are available. An officer who served with 22 SAS in Oman during the civil war in the 1970s recalled that he started with many pieces of private equipment but ended up with a full range of issue kit, simply because he could get replacements and spares through the stores. Nevertheless, the old cliché 'anyone can be uncomfortable' is still valid. Waterproofs are the most obvious example; boots come a close second. In the British Army there is a fairly relaxed attitude to personal equipment, especially since the Falklands conflict. A good example of this is the Norwegian shirt, a cotton jersey garment with a zip-fastening at the neck. It can be worn open or as a roll-neck. First brought back from Norway by the Royal Marines, it is now widely available in British Army unit PRIs (regimental shops).

The private purchase of equipment like bivvy bags has made living in the field easier, but for some soldiers there are weight restrictions on what they carry because they have to carry it on their backs. Camping gas cookers are faster and cleaner than solid-fuel Hexamine or Esbit cookers, but they are bulky. Extras, such as herbs and spices that can be added to issue 24-hour rations, are small and lightweight. One French Army conscript NCO, told me about his squad being inspected prior to an exercise. The men had laid out the contents of their rucksacks: nothing special, just issue clothing and a few comforts. It was when the inspect-

The tank as home. M1 Abrams showing the crew's possessions stowed in the bins and baskets around the turret. Flags and names give tanks an identity.

ing officer reached the kit belonging to a young chef who had been called up that he stopped: the trainee chef was going on exercise with the tools of his future profession, including a pepper grinder, herbs and spices. 'Isn't it heavy?' asked the officer. 'No sir. It gets lighter as we go on,' replied the chef, who kept the squad fed in *cordon bleu* style throughout the exercise.

In greater comfort live the mechanized infantry, armoured troops and any who can prepare a vehicle at their leisure before they take to the woods. Service rations are generally quite palatable if the user takes trouble with them. Thus soldiers in the US Army complain about their MREs (Meals Ready to Eat) which is not surprising, since they appear to eat them cold. The pre-cooked meat meal in a foil sachet is obviously less palatable cold. But put the bag into a container of water and bring it to the boil and the food becomes quite tasty. With the addition of a dash of tabasco, it is excellent. Among the men of 1st Cavalry there was a young soldier who had done an exchange with a British unit, and he was full of praise for the FV432 vehicle. British soldiers who have struggled with old and unreliable 432s may wonder what appealed to a US soldier who is now driving a Bradley – it was the boiling vessel on the rear door.

The US Army conducts its exercises dry: in other words, the canned beer or whisky that makes exercises more tolerable for BAOR is forbidden to the US Army. The Chieftain and Challenger MBTs in service with the British Army have insulated ammunition stowage bins that serve in peacetime as very good cool-boxes for beer. On a sun-drenched Salisbury Plain I still remember the tank troop commander who cheerfully offered me a can of beer – which had a film of condensation on it. It was delicious to a parched

22

FV432, old, in some cases mechanically unreliable, but versatile enough to have welding jobs done that allow kit to be stowed externally. This version has a turret for a 7.6mm machine-gun.

British tank crews break for lunch. They can enjoy German wine with their meals, but US servicemen go into the field 'dry' and watch with envy the way their allies can pack cases of beer or bottles of wine for the exercise.

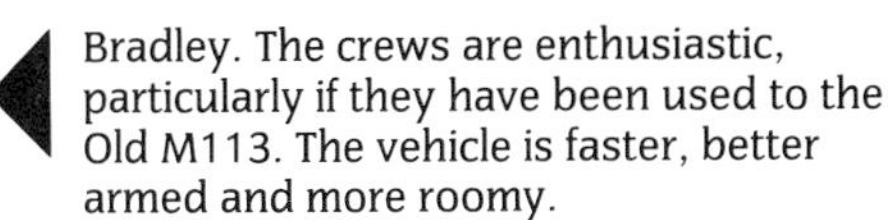

Bradley. The crews are enthusiastic, particularly if they have been used to the Old M113. The vehicle is faster, better armed and more roomy.

Warrior has attracted the same favourable comments from British soldiers who are used to the FV432. It is well armed and has good stowage for ammunition and equipment.

'grunt' who was living out of his water bottle. Alcohol should be treated with care in the field, however. It can be dangerous when temperatures are below freezing or where a soldier has become an exposure case, since spirits expand the blood vessels near the skin surface, and, despite the feeling of warmth they give, they take blood away from the core organs.

Wheels and tracks also allow troops to carry fresh, locally bought rations to supplement their composition ('Compo') rations. A double-burner camping gas cooker is also useful, being faster to light and operate than petrol cookers.

If the soldier can only carry his world on his back or his belt, a simpler cooking kit can include a metal 1944 pattern water bottle or an 85 pattern mug with some strong aluminium foil to cover the top while water is being boiled. Catering-size powdered milk, coffee and sugar can be packed in a pouch together with solid fuel tablets. A 'brew' created with this equipment in the shelter of a wall on a wet and windy day can often taste better than an expensive meal eaten at leisure in a luxury restaurant.

In a close-knit squad or section, the kit to cook in the field can be split among the men just as ammunition, radio equipment or weapons are carried. In the British Army the 'battle partnership', two men who work, eat and fight as a team, is normally the smallest unit.

▲ Above left: An M1 Abrams tank of 1st Cavalry in a non-tactical holding area. Crew possessions are stowed in the packs around the turret. Above: Abrams in the rain. In Germany, exercises come in a season that marks the transition from late summer to early winter.

Exhaust louvres make a useful place to dry clothing. Even when stationary, the M1's power requirements are sufficient to generate enough heat to dry wet garments. ▶

Spare stowage bins and baskets cannot be welded on to the outside of a soldier; what he needs he carries, unless he is vehicle-mounted. Older APCs like the M113 and FV432 had scope for modification. How far this was taken depends on policy sometimes as low as battalion or as high as national army level. The addition of external bins and baskets allows sleeping-bags, water, camouflage nets, fuel and rations to be stowed outside a vehicle, leaving the interior uncluttered. After several days in the field even the most self-disciplined crew of an AFV will begin to be rather smelly. Sweat, fuel, oil and some cordite will combine together to produce a smell that they may not notice, but other people will. The discomfort of life in a tank has not changed much over the years. One British officer emerging grubby from his Chieftain was offered the use of the bath and lavatory by a German householder while on exercise. 'You see,' said the German, 'I was a tank man myself.'

The Germans make some money out of exercises. In 'Certain Strike' rural bread shops were put off-limits to the men of the 1st Cavalry after a company had bought the entire contents of a shop. The soldiers and the shopkeeper were happy, but the locals were none too impressed! I have seen the chance halt by a British Army convoy change the fortunes of a village shop as sweets, biscuits, soft drinks and cigarettes were bought in a rush by the soldiers on the vehicles.

For the men of the US Army III Corps participating in 'Reforger 87' and 'Certain Strike', the headquarters at Fort Hood produced a Soldier's Reference Guide. Similar guides proved their worth when produced for men participating in the 'Bright Star' exercises in Egypt. A little smaller than a paperback, the guide had eighteen chapters, which attempted to provide the user with as much information about Germany as a tourist could wish for while also giving necessary military data such as insignia, command structures and vehicle/aircraft indentification. The guide deserves closer study since it shows a responsible attitude to exercises in foreign countries. Few soldiers will remember a current affairs briefing before an exercise, but they will have time to spare as they obey that military rule of nature, 'hurry up and wait' – then the guide makes excellent reading. (A good paperback is a boon on exercises or operations to fill moments of enforced idleness.)

In the 'Daily living and what an American soldier in Germany needs to know' sections, the reader is faced with shopping, social life and traffic accidents, the

M1 Abrams tanks steer a course around the edge of a field; this avoids damage to standing crops – and advertising their presence to enemy aircraft.

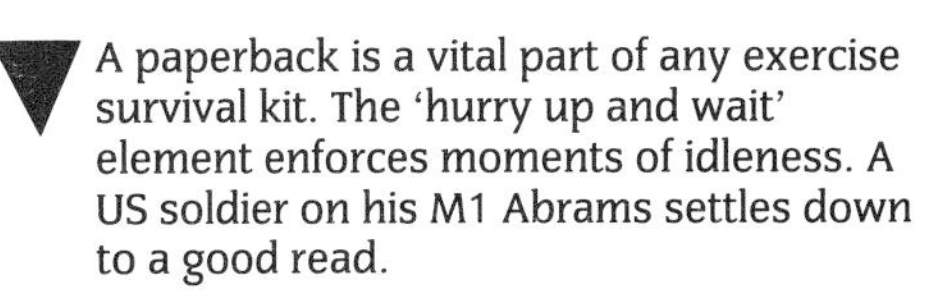

A paperback is a vital part of any exercise survival kit. The 'hurry up and wait' element enforces moments of idleness. A US soldier on his M1 Abrams settles down to a good read.

P

The domestic life of a tank crew. Under their shelter, a crew eat lunch from their ten-man composition (Compo) ration. AFVs always have the advantage that they can carry more rations and comforts than an infantryman, who has to stow his world into a rucksack.

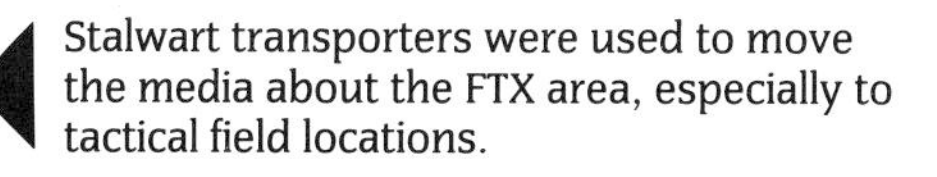

Stalwart transporters were used to move the media about the FTX area, especially to tactical field locations.

1

REFORGER 87... REFORGER 87... REFORGER 87... REFORGER 87...

REFORGER FORGES

NORTHAG

7

NR. 3

4(NL)Div

Orange OBSERVER

SPECIAL NEWSBULLETIN FOR THE ORANGE PARTY DURING THE EXERCISE CERTAIN STRIKE

SPECTACULAIRE LUCHTLANDING BIJ AHLDEN

BRITTEN NEMEN BRU

2

III CORPS & FT HOOD

Public Affairs Office

FACT SHEET

FY 87 #27

21 July 87

Bldg. 60, 287-6906

COMMAND INFORMATION

REFORGER EIGHTY SEVEN

RULES FOR CONTACT WITH MEDIA AND VISITORS

All officially invited media/visitors will be escorted by an escort officer. These officers will be wearing a green armband with the letters P INFO (Public Information) or a white armband with the letters JVB (Joint Visitors Bureau). In addition, the media will carry an Exercise Press Badge.

Only escorted media/visitors have official permission to visit exercise units.

Media without press badges should be asked to register at the APIC (Allied Press Information Center) in the BERGEN HOHNE Training Area at OSTENHOLZ Camp Tel 0 5161 - 7 23 42/7 21 29/7 19 98.

When dealing with visitors do not forget that NATO has invited them and therefore we all should be prepared to act as hosts and behave accordingly.

Talk to media/visitors frankly about your own personal military experience and be prepared to say within security and national limitations:

What you are learning and practicing.
How the exercise is improving your military capability.
How you cooperate with other NATO Allies.
How your equipment operates.

You should remember that you cannot prevent unregistered media or visitors from observing and taking pictures!

Never remove a camera or recording device from anyone!

Call the Military or German Civil Police in the event of a possible security violation.

One important group of visitors will be the CSCE/CDE (Conference on Confidence and Security building measures and disarmament in Europe/Stockholm Agreement) Observers. This group includes observers from NATO and neutral countries as well as Warsaw Pact countries. Questions by observers are to be answered as described above.

DIST: IAW FH Form 1853 (623)

MICHAEL E. MILLER
LTC, SC
Public Affairs Officer

3

REFORGER '87

4

Soviet Military Liaison Mission, Manoeuvre Damage Prevention and even the US Customs on returning to the US. Probably the most unlikely and most imaginative section comes under 'Daily living': the weekend of a German family. It is a small sociological study, contrasting US and West-German lifestyles. In a booklet with details of Warsaw Pact and NATO armies, weapon characteristics and ranges, we find paragraph 5–25: 'For a German housewife, Saturday is the busiest day of the week. The order of the day is hurry, hurry, because normally the shops are only open in the morning . . . .' It may seem a little folksy to some readers, but it must be said that some commercial guidebooks do not have such useful details. Offence can be caused quite innocently by foreign visitors who are unfamiliar with local practices, and the Reference Guide is a wise precaution against this.

I was not offended when I met two US soldiers during the exercise who had taken the contents of Chapter 2, Section V to heart: 'Subversion, espionage and terrorism'. Confronted by an unidentified English-speaking male in civilian clothes with an interest in military matters, they replied to my question: 'I don't know who you are, and I am not going to tell you.'

Operational Security (OPSEC) in the guide includes: 'Avoiding unnecessary discussions of classified material. Discussing information only with authorized persons. Reporting a suspect approach by a foreign agent. Verifying the identity of unescorted visitors to your work area.' And, in what might be a guide to life, 'Choose your friends carefully.'

For both US and British soldiers who have come from home bases, these restrictions can come as a surprise. There is always a local interest in FTXs, even if it is to check on manoeuvre damage, and it can be hard to distinguish the suspect from the innocent enthusiast. Military equipment and weapons are of interest to terrorist groups as well as to Warsaw Pact agencies. Soldiers are constantly reminded of the vulnerability to their equipment – particularly small-

◀ Field training exercises as PR/Propaganda. 1, The Dutch-language *Orange Observer*, a 'Special Newsbulletin for the Orange Party During Exercise Certain Strike'; 2, a guide to US forces from III Corps on how to deal with the Press; 3, '*Reforger 87*', the English-language newspaper for US and British forces; 4, 'Reforger 87' sticker with national flags, NATO symbol and an airborne bridge symbolically linking the USA and West Germany.

▼ A Heavy Expanded Mobility Tactical Truck (HEMTT) brings up fuel to a column of US armour. This 'supply train' unit of the 4th Infantry Division, based at Fort Carson, Colorado, were veterans of the 1985 deployment to Europe.

arms. The danger to civilians from pyrotechnics is another consideration, for children are always keen to play with them.

Living in the field during an exercise, for a visitor or soldier, is a mixture of common sense and prior planning. Thinking of the potential problems before they happen and of the often small ways in which life can be made more comfortable allows the participant to get real value out of the exercises. Two or three weeks in the field can put ordinary life back into context, the 'crises' seem less important when hot water appears straight from the tap and light appears miraculously at the flick of a switch. In many ways the sense of relief with which one greets civlization must be similar to the way in which one would react after experiencing real war.

Radios with the signal codes for the next phase of the exercise are stacked on a vehicle prior to issue.

At the end of the day, an Irish Hussar catches up on the news back home.

His clothes on a 'chairs folding', a British soldier sleeps during the exercise. There is an old saying that goes: 'Don't stand when you can sit, sit when you can lie down, or lie down when you can sleep' – in effect, conserve your energy.

Dutch soldiers perfecting the art of the Dutch military sandwich.

A Bradley crewman bites his way into an MRE freeze-dried fruit course. Some of the food technology in MREs was developed for astronauts' in-flight meals.

A British soldier looks up from his sleeping-bag on the front deck of a Scorpion armoured recce vehicle. Fatigue on exercise as well as night moves means that sleep becomes a priority to be taken when it is available. The 'green maggot' army issue sleeping-bag becomes a temporary haven under these circumstances.

The final survivor – the infantryman with his rifle. Here a British soldier with an SA80 in position beside a Milan ATGW.

Overleaf:
The long and weary road to deterrence and disarmament. M1 Abrams on a road-march through a German wood.

31